We hope you enjoy this book. Please return or renew it by the due date.You can renew it at **www.norfolk.gov.uk/libraries** or by using our free library app. Otherwise you can phone **0344 800 8020** - please have your library card and PIN ready.You can sign up for email reminders too.

PUTTING THE GENIE BACK

Solving the Climate and Energy Dilemma

BY

DAVID HONE

United Kingdom — North America — Japan
India — Malaysia — China

Emerald Publishing Limited
Howard House, Wagon Lane, Bingley BD16 1WA, UK

First edition 2017

Copyright © 2017 Emerald Publishing Limited

Reprints and permissions service
Contact: permissions@emeraldinsight.com

British Library Cataloguing in Publication Data
A catalogue record for this book is available from the British Library

ISBN: 978-1-78714-448-4 (Print)
ISBN: 978-1-78714-447-7 (Online)
ISBN: 978-1-78714-932-8 (Epub)

ISOQAR certified
Management System,
awarded to Emerald
for adherence to
Environmental
standard
ISO 14001:2004.

Certificate Number 1985
ISO 14001

FSC
www.fsc.org
MIX
Paper from
responsible sources
FSC® C013604

INVESTOR IN PEOPLE

' ... *there remains inevitability around political action and legislation to deal with carbon dioxide emissions. It's not just politics that dictates this, but physics. Society can't keep on adding carbon dioxide to the atmosphere and expect nothing to change*'.

CONTENTS

GLOSSARY OF TERMS

AAU	Assigned Amount Unit.
AOSIS	Alliance of Small Island States.
BECCS	Bioenergy Use with Carbon Capture and Storage.
CBDR	Common but Differentiated Responsibilities.
CCS	Carbon Capture and Storage.
CDM	Clean Development Mechanism.
CER	Certified Emissions Reduction.
CFC	Chlorofluorocarbon.
CLG	The Prince of Wales' Corporate Leaders Group (on climate change).
COP	Conference of the Parties
CORSIA	Carbon Offsetting and Reduction Scheme for International Aviation.
CPLC	Carbon Pricing Leadership Coalition.
DACCS	Direct Air Capture of Carbon Dioxide and Storage.
ENSO	El Niño Southern Oscillation.
EOR	Enhanced Oil Recovery.
ERU	Emission Reduction Unit.
ETS	Emissions Trading System.
EU ETS	European Emissions Trading System.

EV	Electric Vehicle (including hydrogen, battery electric and plug-in hybrid vehicles).
GCCSI	Global Carbon Capture and Storage Institute.
GDP	Gross Domestic Product.
GHG	Greenhouse Gas.
HFC	Hydro-fluorocarbon.
HSFO	High Sulphur Fuel Oil.
ICE	Internal Combustion Engine (vehicle).
ICAO	International Civil Aviation Organisation.
IEA	International Energy Agency.
IETA	International Emissions Trading Association.
IGSM	Integrated Global System Modelling.
IPCC	Intergovernmental Panel on Climate Change.
ITL	International Transaction Log.
ITMO	Internationally Transferred Mitigation Outcomes.
JI	Joint Implementation.
LNG	Liquefied Natural Gas.
MBM	Market Based Mechanism.
MITJP	Massachusetts Institute of Technology (MIT) Joint programme on the Science and Policy of Global Change.
NDC	Nationally Determined Contribution (but prefixed with 'I' for intended prior to the adoption of the Paris Agreement).
NDRC	National Development and Reform Commission.

NER   New Entrant Reserve (of the EU ETS).

NET   Negative Emission Technology.

NOAA   National Oceanic and Atmospheric
             Administration.

PV   Photo Voltaic (solar cell).

ROW   Rest of World.

RPK   Revenue Passenger Kilometres.

SLCP   Short Lived Climate Pollutants.

UNEP   United Nations Environment Programme.

UNFCCC   United Nations Framework Convention on
             Climate Change.

WBCSD   World Business Council for Sustainable
             Development.

WMO   World Meteorological Organization.

ENERGY DEFINITIONS AND UNIT ABBREVIATIONS USED

This book uses energy units in discussing the energy system. There are three that are most important.

Joule   The joule, symbol J, is a derived unit of energy in the International System of Units. Approximately 4.2 J is required to heat 1 gram of water by 1°C.

Watt   The SI unit of power, symbol W, equivalent to 1 joule per second, it is the rate of consumption of energy.

Watt-hour   A measure of electrical energy, symbol Wh, equivalent to a power consumption of 1 watt for 1 hour. The Watt-Hour and Joule are interchangeable through a conversion factor of 3600 J/Wh, but Watt-Hour is used for electricity generation, whereas the Joule is used for energy more broadly.

In addition, the energy system is described in terms of primary and final energy.

Primary   Primary energy is an energy form found in nature that has not been subjected to any conversion or transformation process. It is energy contained in raw fuels such as coal, and other forms of energy received as input to a system. Primary energy can be non-renewable or renewable.

Final   Final energy is energy supplied to the final consumer for all energy uses. Electricity is final energy, as is natural gas when used directly for cooking and heating at home. But natural gas can also be classified as primary energy when taken directly from the ground into the energy system for use in a power station.

Barrel   A quantity of oil, 42 US Gallons or about 159 litres. Global oil production is about 95 million barrels per day.

EJ   Exajoules — one quintillion (10^{18}) joules. In 2015 global primary energy use was approximately 550 EJ.

Gt   Gigatonnes — one billion (10^9) tonnes. The largest coal fired power station in the world, a 5.5-GW facility in Taiwan, will emit about 2 billion tonnes of carbon dioxide over its lifetime.

GW   Gigawatt — 1 billion (10^9) watts. A 1-GW power station is the typical size for a modern coal, gas, or nuclear installation. The UK has about 50 GW of installed gas- and coal-fired power-generation capacity.

kg   Kilogramme, an SI unit of mass.

Mt   1 million tonnes.

Mtpa   Million tonnes per annum.

ppm   Parts per million by volume (of a gas in the
           atmosphere).

t   1 tonne or 1000 kg.

TWh   Terrawatt hours − 1 trillion (10^{12}) Watt-hours.
           A 1-GW power station operating for 300 days
           per year will produce about 7 TWh.

°C   Degree Celsius, a measure of temperature.

ABOUT THE AUTHOR

David Hone is Chief Climate Change Advisor at Shell International Ltd. He joined Shell in 1980 after graduating as a Chemical Engineer from the University of Adelaide in Australia, and previously held positions in refinery technology, oil trading, and shipping areas for Shell. David has been the principal climate change adviser for Shell since 2001 and has represented the company in that capacity in a wide variety of forums. He is a board member of the International Emissions Trading Association (IETA), was Chairman of IETA from 2011 to 2013, and is a board member of the Center for Climate and Energy Solutions (C2ES) in Washington.

FOREWORD: DONALD TRUMP AND THE PARIS AGREEMENT

As this book was being readied for publication, President Donald Trump announced that the United States would withdraw from the Paris Agreement. Heads of State from all corners of the world quickly responded, vowing to uphold the Agreement and ensure its continuance. But can the Paris Agreement, as discussed extensively in the pages that follow, survive?

There is an element of déjà vu to this event. Just days into my new job as climate change adviser in Shell, then President George W. Bush announced that the United States was withdrawing completely from the Kyoto Protocol and would follow an alternative path forward in terms of climate action. At the time, he proposed a significant step-up in technology development through the National Climate Change Technology Initiative and a leadership role by the United States to work within the United Nations framework and elsewhere to develop with its friends and allies and nations throughout the world an effective and science-based response to the issue of global warming.

The June 2001 Bush announcement was widely expected and indeed, it helped spell the end for the Kyoto Protocol. The UNFCCC process fractured as a result, with some Parties continuing to pursue the Kyoto

Protocol and all Parties brought back to the table to negotiate a new deal that worked for the United States. This gave birth to the Ad-Hoc Working Group on Long Term Cooperative Action, which together with the Kyoto Protocol arrangements could have potentially combined into a satisfactory global deal. Unfortunately, they didn't, with the meltdown in Copenhagen being the outcome. But the pieces were reassembled, in large part led by the United States under President Obama, with the result being the Paris Agreement in December 2015. That process took over 14 years to complete.

Sixteen years on from the Bush announcement and again from the White House lawn, President Trump has now declared that the United Stated will exit the Paris Agreement, once again with the caveat that the Administration would be open to a renegotiation or even an entirely new agreement. The reasons given are largely the same as those of President Bush; unfairness, competitiveness concerns, negative economic impact, layoffs of workers and price increases for consumers. But the circumstances are very different this time around.

Looking 14 years ahead from today we will be in the 2030s. My own analysis in Chapter 2 shows this as the time when we may start to see years in which the global average temperature rise could equal or exceed 1.5°C above the level in the mid-1800s. There isn't any grace period remaining to reorganise, negotiate and agree yet another climate deal. Nor should that happen; the Paris Agreement is structured to reflect what countries can offer, with no requirement other than that successive offers should improve over time and ratchet towards the end goal of net-zero emissions in the second half of the century. The Agreement isn't a good or bad deal for anyone; it simply reflects the progression required over time as nations either continue or begin to proactively manage emissions and eventually contain them.

The Paris Agreement is made up of national contributions, determined by nations per their domestic circumstances. This is the case for all countries, from the United States of America as the world's largest economy through to Zimbabwe as one of the poorest. Although the Agreement asks for developed countries to implement economy-wide absolute emission reduction targets, there is an expectation that all countries move in this direction and the Agreement encourages such movement. Several developing countries have structured their national contributions to reflect this and in the time since the negotiations concluded, more have implemented measures to that effect. For example, China is implementing a nationwide emissions trading system and emissions within their economy are now expected to peak well before 2030, ahead of their stated national contribution.

With all nations supposedly on a pathway towards absolute targets and eventually net-zero emissions, there is nothing left to negotiate other than the timeline along which this proceeds, although as I discuss in Chapter 3 the current timeline leaves the world well short of a 2°C outcome. Once again, the Agreement sets out the process for addressing this, rather than Parties having to resort to yet another negotiating process towards an alternative agreement. There is a transparency framework, a stocktake process and a mechanism to facilitate implementation of and promote compliance with the provisions of the Agreement. Although the proposed mechanism is facilitative in nature and should function in a manner that is transparent, non-adversarial and non-punitive, it nevertheless offers the opportunity for a country such as the United States to negotiate more rapid convergence of effort.

Both Chancellor Merkel and President Macron, along with UNFCCC Executive Secretary Patricia Espinosa, made it clear the morning after President Trump's announcement that there would be no renegotiation of the Paris Agreement. While anything is possible in theory, a renegotiation would put an end to the Agreement and probably not deliver a replacement for a decade or more. They were right to reject the proposal; in any case, it simply isn't necessary under the structure that exists.

Given all the above, the current Administration may still be concerned about the effort required by the United States to deliver its stated goal of a reduction of 26–28% in emissions by 2025 against a 2005 baseline, particularly when compared to some countries. Although the current surge in natural gas production and its replacement of coal for power generation, the advance of renewable energy and the roll-out of electric vehicles are all contributing to a fall in US emissions, the target remains ambitious. While the prospect of success is visible within the energy transition that is underway, the United States could simply resubmit its national contribution. Various parts of the Paris Agreement and the accompanying Decision Text open the door to such a step, and former Secretary of State John Kerry, who negotiated the Agreement for the United States, said as much on the BBC shortly after the Trump announcement. Although successive national contributions are required to demonstrate increased ambition under the Agreement, this is the first such submission and therefore can be revised.

By resubmitting its national contribution, some semblance of renegotiation would be achieved, at least in part. A new contribution from the United States would still require an absolute target as this is required of developed countries under Article 4.4, but the number could have a much wider margin, covering the expectation of economic growth that the President alluded to in his speech on June 1. Of course, this isn't an ideal outcome from an emissions perspective, but it would keep the United States in the frame for some time to come and allow them to pursue further equivalency of effort through the implementation mechanism.

Should the United States end up on a path of true departure, this will still take until November 2020 to execute. A Party cannot serve notice of termination until three years after the Agreement enters into force and then there is a period of one year before their participation ends. While such a period does not extend beyond the term of the current Administration, it nevertheless represents a long time in politics.

In the meantime, some 195 other countries will continue to implement their national contributions through a variety of approaches. The European Union, for example, is pursuing a reduction of 40% by 2030 against a 1990 baseline, utilising a cap-and-trade system for the large emission sources such as power stations. Even within the United States, the current energy transition will continue, with much the same result in terms of emissions in 2020 and possibly even 2025 as would have been the case with the national contribution in place. Deactivation of the contribution is unlikely to spur new construction of coal-fired power stations given the intense competitive pressure from natural gas and renewables. Even discounting current competition, there remains the prospect of future carbon constraints imposed at some point within the 50+ year lifetime of a new coal-fired power station.

But the energy transition is just one element of the Paris Agreement; emissions management is at its core. This will require more than just an energy transition to implement, probably requiring large-scale deployment of negative emissions technologies which I discuss at length in Chapters 3, 5 and 7, including geological storage of carbon dioxide. This latter step may be the one that suffers following the US announcement.

The Paris Agreement can and likely will survive the events of June 1. But if other nations don't step up and look beyond their own energy transitions, focussing squarely on the need for a net-zero emissions outcome within the next 50-80 years, then the goal of the Agreement may be at risk.

INTRODUCTION

In mid-2008, the head of the Shell media team dropped by my desk with a proposal for the company to take an early step out of the world of traditional corporate communication and into the then new and emerging world of social media. The idea was to set up a regular blog series that discussed issues pertinent to the company and its stakeholders. With climate change a central issue for society, the plan was to start on this subject. As the leading climate change person within the company, I was asked to think about topics to kick off the initiative. A few months later I was up and running with my first posts covering emissions trading, policy development, and the energy transition. The opportunity was also a great fit with my role, which requires me to be something of an independent voice internally on climate change.

The blog is now heading towards a decade of posts and somewhat perversely, it has rather outlasted the initial enthusiasm for the idea. By 2017 there were nearly 400 posts, with several hundred thousand words of content covering almost every aspect of the climate issue. I have found that readership is quite wide, mainly through direct feedback from readers who I meet by chance at conferences and even socially.

As a chemical engineer with 37 years' experience in the oil and gas industry, my goal has always been to tackle the climate issue from an engineering perspective; based on data, built on facts and without the histrionics and emotion that have come to define this subject in many quarters. In 2014, I began work on a series of three short e-books that bought to life some of the ideas from my blog, succinctly covering many of the pertinent issues of climate change today, including carbon trading and the Paris Agreement.

This book brings together and builds on my blog and the e-books. It tells the story of the climate change issue and the transition in the energy system that must be implemented to finally address the issue. At its most ambitious, the Paris Agreement implies economic and societal change on a scale that sees carbon dioxide emissions fall rapidly from 40 billion tonnes per annum in 2016, to net-zero by the middle of the century. Yet our fossil fuel based energy system which ushered in the Industrial Revolution nearly 200 years ago continues to grow and evolve even as new sources of energy come into the market and compete.

The principal economic instrument for change is clear and has been for over two decades, but in 2017 only a fraction of the global economy actively employs government led carbon pricing policies and within that none of these systems operate at a level commensurate with the pace of change that is necessary. As deployment of new energy technologies accelerates, can solutions be found to cover the full range of services delivered by fossil fuels and can warming be limited to the agreed global goals? The book explores the climate issue from its very beginnings through to the end of the 21st century and looks in depth at the transition challenge that society faces.

Data from the book are sourced from Shell and from the University of Oxford, IEA, NASA, NOAA and CDIAC and all proceeds of the book will go to an NGO working on climate change-related issues.

1

ENERGY AND CLIMATE CHANGE

In July 1912, in rural New South Wales, Australia, the *Braidwood Dispatch and Mining Journal* published a short article on the global use of coal. The same article appeared a month later in a similar local publication in New Zealand.

COAL CONSUMPTION AFFECTING CLIMATE

The furnaces of the world are now burning about 2,000,000,000 tons of coal a year. When this is burned, uniting with oxygen, it adds about 7,000,000,000 tons of carbon dioxide to the atmosphere yearly. This tends to make the air a more effective blanket for the earth and to raise its temperature. The effect may be considerable in a few centuries.

These stories were extracted from a longer article published in *Popular Mechanics* in March 1912. That article commented on the extreme weather of 1911 and drew on the late 19th-century work of Svante Arrhenius, a Swedish chemist who linked the average surface temperature of the Earth to the level of carbon dioxide in the atmosphere. From 1896 when Arrhenius began publishing his work, a number of similar stories appeared. At the time many cities, but famously New York and London, were also seeing the local impact of coal burning.

Just over 103 years later, when French foreign minister Laurent Fabius banged his gavel on the evening of Saturday 12 December 2015 in Paris, he ushered in a truly global deal on climate change. It embraces the spirit and ambition necessary to finally deal with the very issue that the *Braidwood* editor had noted with regards to the use of coal, but in the years that followed included all use of fossil fuels and many other practices in what has now come to be known as the Anthropocene[1] era. Less than a year later, on 4 November 2016, the Paris Agreement entered into force after the ratification criteria had been met.

But will the Paris Agreement really see us through to the end of this century, bringing an end to anthropogenic[2] emissions of greenhouse gases (GHG) and therefore limiting warming of the climate system?

The track record for confronting the climate issue is not good. The 21st Conference of the Parties (COP/COP21) in Paris also marked the last opportunity for most of us to breathe fresh air with a carbon dioxide level below 400 parts per million (ppm), compared with a level of 275 ppm before the start of the Industrial Revolution. This represented a near 50% increase in atmospheric carbon dioxide in less than 200 years. The first full day of 400 ppm carbon dioxide as recorded at the Mauna Loa Observatory in Hawaii was in May 2013, but because of the annual vegetation cycle that impacts the level of carbon dioxide in the atmosphere, there was still time to enjoy the heady days of the three hundreds. This date and the more recent last day at 399.9 ppm produced an outpouring of sentiment and grief from some environment correspondents, but the news has seemingly been forgotten.

From a human perspective, 400 ppm is little different from 300 or 500 ppm and in any case the level of carbon dioxide in a cramped meeting room can be much higher, but there is significance in the number nevertheless. The background level of carbon dioxide in the atmosphere moves slowly over millennia in response to gradual changes in ocean temperature, release from volcanic activity and uptake through weathering. The amount of carbon dioxide present in the atmosphere at any point in time is an important determinant in establishing the surface temperature of the planet, with higher levels of carbon dioxide linked to a higher temperature.

That relationship was established in the late 19th century and widely reported on after Svante Arrhenius published his landmark paper 'On the Influence of Carbonic Acid in the Air upon the Temperature of the Ground'. Since the beginnings of collective agriculture and more recently with the onset of the Industrial Revolution, the level of carbon dioxide in the atmosphere has been rising. In 1958 Charles Keeling started collecting samples of air at Mauna Loa Observatory in Hawaii, which led to both the development of an instrument to accurately measure atmospheric carbon dioxide and the publication of the modern time series of carbon dioxide observations. Keeling found that the time series was rising year on year, but also fluctuated within a year as the northern hemisphere moved through the growing seasons.

Unfortunately, the arrival of this 400 ppm day had become inevitable. Since the early days of the Keeling Curve at 315 ppm, when it became clearly apparent that anthropogenic carbon dioxide emissions were accumulating in the atmosphere, the ppm have been counting up.

Before the start of the Industrial Revolution and as the modern coal era was just getting going, the level of carbon dioxide in the atmosphere was around 275 ppm. It had been at a similar level for some 10,000 years prior and fluctuated between 175 and 300 ppm over the previous million years, both as a driver of and in response to warmer and cooler climates.

So began the Industrial Revolution with James Watt inventing the steam engine at 278 ppm. By 300 ppm the internal combustion engine was powering the first tanks at the Battle of the Somme.

Despite a very clear recognition of the potential impact of rising carbon dioxide levels coming in the form of a White House report from the President's Science Advisory Committee during the Johnson Administration at 321 ppm, it wasn't long before there was a brief worry about global cooling. Then, with atmospheric chemistry growing as a discipline (probably on the back of concerns about a Cold War—induced nuclear winter), society was distracted at 332 ppm by the first major anthropogenic global concern, the hole in the ozone layer. But with a treaty negotiated and ratification underway by 349 ppm (only 17 ppm to sort that one out), it didn't take long for the science community to remember that another big issue was lurking in the shadows.

At 352 ppm and nearly 40 ppm on from the start of the Keeling Curve, NASA climate scientist James Hansen stated to a US Congressional Committee under the presidency of Ronald Reagan that the Earth was warmer in 1988 than at any time in the history of instrumental measurements. He argued that global warming is now significant enough that the greenhouse effect can be ascribed, with a high degree of confidence, as a cause and effect relationship and that computer simulation indicates that this enhanced greenhouse effect is already large enough to begin to affect the probability of extreme events such as summer heat waves.

The international diplomatic process that led to the Paris Agreement had taken 25 years (48 ppm), starting at the Second World Climate Conference in Geneva from 29 October to 7 November 1990. Sponsored by the World Meteorological Organization (WMO), the United Nations Environment Programme (UNEP) and other international organisations, the main objectives were to review the UNEP/WMO World Climate Programme and to recommend policy actions. The participants called for elaboration of a framework treaty on climate change and the necessary protocols, containing real commitments and innovative solutions, in time for adoption by the UN Conference on Environment and Development in June 1992. The latter event came to be known as the Rio Earth Summit and it did indeed result in the adoption of the United Nations Framework Convention on Climate Change (UNFCCC), which set out a framework for action aimed at stabilising atmospheric concentrations of GHG to 'prevent dangerous anthropogenic interference with the climate system'.

But it was to be another 13 ppm on from the Hansen intervention before the Kyoto Protocol was adopted by parties to the UNFCCC and 14 ppm more before it was finally ratified at 380 ppm. That Protocol placed emission reduction obligations on developed countries and encouraged developing countries towards a cleaner energy future through an incentive mechanism linked to the goals adopted by the developed countries. 21 ppm later and with 400 ppm now in the rear view mirror, the Kyoto Protocol is a shadow of its former self, but with at least the legacy of the origins of a global carbon market. In the interim there was a valiant attempt at a new global deal in Copenhagen, although even that was 14 ppm before COP21.

The Paris Agreement sets out a goal to stay well below 2°C, so likely below 450 ppm, which on the face of it looks very ambitious, given that the atmospheric concentration was already at 400 ppm as the doors to the Le Bourget Conference Centre closed. But a lot can happen in 50 ppm; the first World Wide Web page posted in 1993 was only 43 ppm prior to COP21 in Paris.

As COP21 concluded, I was reminded of a quote by Otto von Bismarck, 'Laws are like sausages, it is better not to see them being made'. Yet, over the course of many years I had done just that. I could now reflect upon the complex and torturous course of modern diplomacy that had worked to deliver a deal and which hopefully represents renewed global leadership on climate change.

Having spent my entire career in the energy industry, I've had the opportunity to see and experience fossil fuel production and consumption in all its many forms. As a chemical engineering student I had the chance to work for a largely coal-based electricity company in Australia and briefly in a cement plant but on completion of my degree in 1980 I started with Shell, which is where I've remained until today.

Working in refineries, managing oil tankers and visiting Shell locations such as the Alberta Oil Sands has provided me with hands-on experience with the hardware involved in the energy industry. However, it was my 10 years in the world of crude oil trading and shipping that offered real insight into the size and complexity of the system society has collectively built to power the world. The Shell trading room in London is just one of many such trading centres scattered around the globe. Despite my refining experience, I was staggered by the ceaseless activity of buying and selling crude oil, chartering ships, discharging cargoes, liaising with loading terminals and transferring hundreds of millions of dollars in cargo payments.

Why then, in 2001 with over 20 years of experience in the oil industry, did I venture into the climate change debate? I decided to take up what was then the only job[3] in Shell dedicated exclusively to the climate issue, created some three years earlier in the wake of the agreement of the Kyoto Protocol.[4] My selling points were an industry background, commercial experience and a good knowledge of energy trading markets, which at the time aligned well with the aim of moving towards a globally traded carbon market. I also had a personal interest in environmental issues that were global in scale, which had started back in high school with an article I wrote for our science journal on the impact chlorofluorocarbons (CFC[5]) were having on the ozone layer. The fact that the ozone layer problem had been addressed and CFC production drastically curtailed as a result of the Montreal Protocol — the

1989 international treaty designed to protect the atmosphere by progressively eliminating substances responsible for ozone depletion — gave many climate-issue advocates hope that a similar deal could be reached for carbon dioxide emissions. But perhaps they had never stopped to meet the energy system.

In 2002, CEO of BP, Lord John Browne, gave a landmark presentation on climate change mitigation in the City of London that I was fortunate to attend. He took the opportunity to introduce the idea of stabilisation (or reduction) wedges to a mainly City audience (the work had already circulated in the academic sector). Stephen Pacala and Robert Socolow at Princeton had developed the wedge idea in a research programme supported by BP. Each wedge represented one of a number of quantifiable actions that when combined together were necessary to move from a business-as-usual global emissions trajectory to a particular atmospheric stabilisation of carbon dioxide. In the initial study that stabilisation was 500 ppm, well above what is now considered as the level not to breach.

Reduction wedges were on a very large scale (up to 1 Gt carbon/annum or nearly 4 Gt carbon dioxide) and consisted of actions such as:

- Increasing the fuel economy for 2 billion cars from 30 to 60 miles per gallon;

- Replacing 1400 1 GW 50%-efficient coal plants with gas plants (four times the then production of gas-based power);

- Introducing Carbon Capture and Storage (CCS) at 800 1 GW coal-fired or 1600 1 GW natural gas (compared with 1060 GW coal operating globally in 1999) power plants;

- Adding 700 GW (twice the year 2000 capacity) of nuclear fission capacity.

At least a dozen reduction wedges were proposed of which four examples are given above. The task confronting the world was to implement all of them if anticipated energy needs were to be met and carbon dioxide emissions reduced; it wasn't a case of picking and choosing.

Lord Browne fully understood the issue and was trying hard to convey the enormity of the mitigation task. This was the first real attempt to quantify the physical changes in the energy system that were necessary and to turn them into a popular narrative. Many variations on the approach followed in subsequent years. Yet well over a decade later, incredibly, none of these proposals have been put into practice. The current growth in solar photovoltaic (PV) perhaps comes closest.

More recently, researchers from universities in the United States and China re-examined the wedges and concluded that the scale of the issue had grown and that an even more ambitious set of wedges would be required to address the climate issue. The team behind this analysis introduced the concept of 'phase-out' wedges, or wedges that represent the complete transition from energy infrastructure and land-use practices that emit carbon dioxide (on a net basis) to the atmosphere to infrastructure and practices that do not.

Understanding the climate issue today isn't just about recognising that atmospheric carbon dioxide levels are rising and therefore emissions must be reduced. While this is true, it oversimplifies a complex problem which now involves two huge and unwieldy beasts, the ocean/atmosphere system and the energy system. Neither is easily changed in a short space of time. The problem that society faces is the clash between the two. Our all-consuming demand for energy means that a delivery and use system has been built on a scale equivalent to the atmosphere itself, giving the former the capacity to affect the latter. And that is what is happening.

The scale of the global energy system isn't easy to grasp. Perhaps you need to see it. After four years studying chemical engineering, I took a few months off to travel to the United Kingdom and United States. In 1980 the flight from Melbourne to London involved a stop in Perth, then a second stop in Bombay (as it was then called), after which there was a final ten-hour leg to complete the journey. I recall the plane leaving Bombay at midnight with clear skies. It wasn't long before a string of brilliant orange beads of light appeared on the horizon. Given that we were over the Indian Ocean, with only water ahead of us, I was perplexed as to what this might be. For nearly an hour, perhaps two, the lights hovered in the distance, growing steadily more intense, without an obvious source.

The flight path took the plane over Oman and directly up the western coast of the Persian Gulf, at which point the source of the lights became apparent. These were gas flares coming from offshore oilrigs, oil terminals, refineries and chemical plants. The air was crystal clear. Such was the intensity of the light that even from 30,000 feet they produced an orange glow as fierce as the evening sun. Every detail of every installation for mile upon mile was visible.

This was the oil industry, as it then existed, where gas was an inconvenient by-product without a local market and the Liquefied Natural Gas (LNG)[6] industry that can take this by-product today was in its infancy. The industry has significantly reduced the practice of flaring natural gas that is brought to the surface with crude oil production, but the scale of the oil industry has only increased. Between 1980 and 2016 global oil production grew by about 50%.

Now, obviously, not everyone has the opportunity to witness large-scale energy production first-hand, so perhaps a few examples will help. In the two hours that you might spend watching the Leonardo DiCaprio climate change movie *Before the Flood*, a lot will happen in the world. It's become a very busy place powered by a lot of energy. Just to keep up with current energy demand, the next two hours will see:

- Four Very Large Crude Carriers (VLCC) of oil loaded somewhere in the world. In total, that's more than enough oil to fill the Empire State Building.

- About 2 million tonnes of coal extracted. Much of this moves by rail and if all this coal were carried in a single train, that train would be about 200 miles long.

- 800 million cubic metres of natural gas produced, which under normal atmospheric conditions would cover the area enclosed by London's M25 to a depth of about a foot; after half a day everyone in London would be breathing natural gas.

- 8–10 cubic kilometres of water passing through hydroelectricity stations, or enough water to more than fill Loch Ness.

Our immediate contact with this is the fuel for our cars, the electricity that lights our homes and powers our stuff and the oil or natural gas used in our boilers. But there is more, much more. This includes the unappealing, somewhat messy but nevertheless essential chemical plants where products such as sulphuric acid, ammonia, caustic soda and chlorine are made (to name but a few). Combined, about half a billion tonnes of these four products are produced annually. Manufactured by energy-intensive processes operating on an industrial scale, but concealed from daily life, these four products play a part in the manufacture of almost everything. Even the ubiquitous can of soft drink relies on sulphuric acid; the chemical is used to give the aluminium can the shiny look that is expected before opening it and consuming the contents. These core base chemicals rely in turn on various feedstocks. Sulphuric acid, for example, is made from the sulphur found in oil and gas and removed during refining and treatment processes. Although there are other viable sources of sulphur, they have long been abandoned for economic reasons.

Then there is the stuff we make and buy. The ubiquitous mobile phone and the much-talked-about solar PV cell are just the tip of a vast energy-consuming industrial system that relies on base chemicals such as chlorine, but also materials such as steel, aluminium, nickel, chromium, glass and plastics from which the products are made. The production of these materials alone exceeds 2 billion tonnes annually. All of this is made in facilities with concrete foundations, using some of the 3−4 billion tonnes of cement that is produced annually. The energy system transition that society is looking to as a solution to the rising level of carbon dioxide in the atmosphere is also dependent on these materials. The current formulation for the battery that powers a 2016 model Tesla includes some 35 kg of nickel. Scaling production to the extent that electric vehicles (EV) might quickly put paid to gasoline and diesel powered vehicles implies a rapid doubling in global nickel production, which also means more energy use.

The global industry for plastics is also rooted in the oil and gas industry. The big six plastics[7] all start their lives in refineries as base chemicals extracted from crude oil.

All of these processes are energy intensive, requiring gigawatt[8]-scale electricity generation, high-temperature furnaces and large quantities of high-pressure steam to drive big conversion reactors. The raw materials for much of this come from remote mines, another hidden key to modern life. These, in turn, are powered by utility-scale facilities, huge draglines for digging and 3 kilometre-long trains for moving the extracted ores. An iron ore train in Australia might be made up of 300–400 rail cars, moving up to 50,000 tonnes of iron ore, utilising six to eight locomotives. These locomotives run on diesel fuel, although many in the world run on electric systems at high voltage, e.g. the 25 kilovolt AC iron ore train from Russia to Finland.

This is just the beginning of the energy and industrial world we live in which is largely powered by the use and combustion of oil, gas and coal. These bring economies of scale to everything we do and use, whether we like it or not. Not even mentioned above is the agricultural world that now feeds over 7 billion people, uses huge amounts of energy and requires its own set of petrochemical-derived fertilizers and pesticides. The advent of technologies such as 3D Printing may shift some manufacturing to small local facilities, but even the material poured into the tanks feeding that 3D machine will probably rely on sulphuric acid somewhere in the production chain.

Despite all this, over 1 billion people in the world today have little or no access to modern energy services. Another 2 billion or so have just climbed onto the bottom rung of the energy ladder and will begin to consume in order to raise their living standards.

A related question is whether a country can develop without an accessible resource base of some description, but particularly an energy resource base. A few have done so, notably Japan (although even in that country coal mining rose to some 55 million tonnes per annum during the Second World War), but most economies have developed on the back of coal, oil, gas and minerals. It was the use of coal that supported the rise of industry in Germany, Great Britain, the United States and Australia and more recently in China, South Africa and now India. Of course, strong governance and institutional capacity are also required to ensure widespread societal benefit as the resource is extracted.

Coal is a relatively easy resource to tap into and make use of. It requires little technology to get going but offers a great deal, such as electricity, railways (in the early days), heating, industry and very importantly, smelting (e.g. steel making). For both Great Britain and the United States, coal provided the impetus for the Industrial Revolution. In the case of the latter, very easy-to-access oil soon followed, and mobility flourished, which added enormously to the development of the continent.

But the legacy that this leaves, apart from a wealthy society, is a lock-in of the resource on which the society was built. So much infrastructure is constructed on the back of the resource that it becomes almost impossible to replace or do without, particularly if the resource is still providing value.

As developing economies emerge, they too look at resources such as coal. Although natural gas is cleaner and offers many environmental benefits over coal (including lower carbon dioxide emissions), it also requires a higher level of infrastructure and technology to access and use, so it may not be a natural starting point. It often comes later, but in many instances it has been used as well as the coal rather than instead of it. Even in the United States, the recent natural gas boom has not completely displaced its energy equivalent in coal extraction; rather, some of the coal has shifted to the export market.

On the back of all this rests the issue of rising levels of carbon dioxide in the atmosphere and warming of the climate system.

2

CARBON DIOXIDE EMISSIONS, TEMPERATURE AND GLOBAL CHANGE

CARBON DIOXIDE AND TEMPERATURE

Carbon dioxide is a greenhouse gas in that it causes warming of the planet's surface above what it would otherwise be. A higher concentration in the atmosphere leads to additional warming of the surface until the equilibrium heat balance is restored. This is rooted in physics — so why has this type of warming become a contentious issue? Why are lawyers and politicians now arguing about physics?

Nobel laureate in chemistry and Mexican climate change policy advisor Mario Molina is best known for his work in identifying the role of CFCs in the destruction of the ozone layer. At a June 2011 forum for the MIT Joint Program on the Science and Policy of Global Change (MITJP) where experts on climate science come to discuss key developments, Molina's keynote address took up the issue of how to convey an understanding of climate change science to the general public. He lamented the inability of scientists to get their message across to the public regarding what is, in his view, a relatively simple and well-understood physical phenomenon governed by a set of equations. He put the question to the audience, "What is it about Planck's Law and the Boltzmann constant that is now in dispute?" A similar question was asked regarding Kirchhoff's Law and the other equations used to calculate the observed temperature of the atmosphere,[9] all of which have been around since the 19th century. Most if not all of these physical laws were discovered for reasons unrelated to atmospheric chemistry, but through the development of disciplines such as quantum theory. However, applied to the discipline of atmospheric chemistry, they do much to explain the physical phenomena that surround us in the world.

The above laws and constants have been proven in practice many times over — if not then the devices we use in everyday life, from iPads to microwave ovens, wouldn't work, nor would they exist in the first place. All depend on the same physical and quantum principles that make up our understanding of the workings of the atmosphere and the impact of a change in its composition.

Professor Molina didn't have a solution to the difficulty of communicating climate science other than to remind us of the initial scepticism towards CFCs, followed by action and international agreement as to how to tackle the problem of their emission into the atmosphere. He noted that this was, to some extent, down to the role of the business sector and the introduction of new refrigerants, developed to replace CFCs. Unfortunately, the climate problem is far more complex than the depletion of the ozone layer, given our reliance on fuels and industrial processes.

A good start would be to recognise that physics and chemistry govern our lives and that the society we have built depends on the laws, constants and algorithms developed from these disciplines, which includes our understanding of the processes in the atmosphere.

The impact of emissions on a given ecosystem, be it chemicals into water or carbon dioxide into the atmosphere, can be described in one of two ways.

i. If the material being emitted remains in the environment for a short while before it breaks down, is deposited somewhere or leaves with the main flow through the system (e.g. river water), then the impact that it has is largely related to the rate at which it flows into that system at any given time. For example, a chemical which happens to degrade very quickly that is slowly dripping into a lake may never cause a problem as its rate of degradation ensures that the concentration of the chemical never rises to a toxic level, at least until the rate of discharge exceeds the rate of degradation. Even then, once the discharge stops, the problem quickly rectifies itself. This is a flow problem and the rate at which the material is emitted on a daily, weekly or yearly basis is all-important.

ii. If the material is very slow to be removed and doesn't break down, it will then tend to accumulate and its impact will grow and grow. Even a very small discharge will eventually cause a problem. If the emission finally stops the problem may at least stop getting worse, but it won't get any better until the material is removed, either through some natural process or by intervention. This is a stock problem and the key determinant here is the total amount of emissions over time. The instantaneous rate of emissions is far less important.

What can be seen, as fossil fuels are used, is an ongoing flow of carbon dioxide into the atmosphere from the combustion process. Armies of statisticians produce pages of data on the rate at which this is occurring, typically on an annual basis. The International Energy Agency (IEA) publishes an annual assessment of carbon dioxide emissions, detailing the output according to country, industry, fuel type, region and use, slicing and dicing the data. However, the words 'rate' and 'flow' don't correspond to the way in which the atmosphere responds to societal emissions of carbon dioxide. Rather, the anthropogenic emission of carbon dioxide to the atmosphere is a stock problem, meaning the total emissions over the whole past and future industrial era is what matters.

Looking at the carbon dioxide emissions problem, imagine the Earth's surface consisting of three parts.

i. The geosphere goes down from the surface to a depth of several kilometres, where some 4 trillion tonnes (at least) of carbon is stored as coal, oil and natural gas. There is also an enormous quantity of limestone, a carbonate.

ii. The biosphere is primarily the land where carbon exists in trees, plants, soil and living things.

iii. The ocean/atmosphere system is where carbon exists as carbon dioxide either dissolved in the ocean or as a trace gas in the atmosphere.

Over time, a carbon equilibrium has been established between these systems, with the level in the atmosphere plateauing for thousands of years at a time, despite the continuing transfer of carbon between the three parts. We also know that the surface temperature of the planet is related to the level of carbon dioxide in the atmosphere. This relationship was established by physicist John Tyndall in 1864 and chemist Svante Arrhenius in 1896. Long before vested interest groups, lobbyists and environmental NGOs argued about global warming, Arrhenius had calculated that if atmospheric carbon dioxide were to double in concentration the global temperature would rise by some 5°C.

The amount of carbon dioxide in the atmosphere does fluctuate, but always as a result of some physical change in the environment. For example, during the last ice age, the amount of solar energy reaching the planet dipped due to orbital variations of the Earth around the sun and, consequently, the ocean/atmosphere system cooled. This allowed the amount of carbon dioxide dissolved in the ocean to increase and therefore the atmospheric carbon dioxide level dropped. That drop resulted in further cooling, causing the system to shift until a new equilibrium was reached. Given the size of the system and the very slow rate at which the equilibrium shifts, these changes would typically take tens of thousands of years.

But now a shift is underway taking just decades, not millennia.

Three types of anthropogenic activity result in a rapid shift in carbon, disturbing the equilibrium:

i. Our use of fossil fuels. As coal, oil and natural gas are extracted and used, stored carbon from the geosphere is being rapidly shifted to the ocean/atmosphere system as carbon dioxide.

ii. Our use of fossil limestone to make cement. Heating limestone releases carbon dioxide into the atmosphere.

iii. Land use change. As the planet is deforested and the upper soil layer disturbed for agriculture and urbanisation, stored carbon from the biosphere is released into the ocean/atmosphere system.

The available removal mechanism is very slow — thus carbon dioxide is accumulating in the ocean/atmosphere system at a much faster rate than it is being removed. The difference is several orders of magnitude when compared to its return to the geosphere through processes such as weathering and ocean sedimentation. This situation is akin to the stock problem described earlier. What really matters and therefore where the focus needs to be is the cumulative amount of carbon dioxide released over time from fossil sources and land use change.

Over the period from 1750 to 2016, i.e. the entire industrial era, approximately 610 billion tonnes of fossil and land-fixed carbon was released and entered the atmosphere (as 2 trillion tonnes of carbon dioxide[10]). About half of this carbon dioxide dissolved relatively quickly in the ocean or was absorbed into the land-based biosphere, while the remainder stayed in the atmosphere. As a result, the concentration of carbon dioxide in the atmosphere rose from 275 ppm in 1750 to 400 ppm globally in 2016. The rate at which the concentration is rising is also rising, moving from 1 ppm per annum in 1960 to nearly 3 ppm per annum in the first year after the Paris Agreement. This process is shifting the carbon equilibrium that has existed since the start of the current interglacial warm period (the Holocene). In turn, this will have an observable impact on other features of the same equilibrium, including the surface temperature of the planet.

Since the days of Arrhenius, a great deal of work has been done on the relationship between the level of carbon dioxide in the atmosphere and the resulting temperature change. More recently, the focus has been on how much the temperature will rise for a given cumulative release of carbon dioxide over time.

This work has been pioneered by, amongst others, Myles Allen, a physicist at Oxford University. I first met Myles, Professor of Geosystem Science in the School of Geography and the Environment, and Head of the Climate Dynamics Group in the University's Department of Physics, at a symposium held by the MITJP in 2003. In a landmark paper,[11] Professor Allen and his colleagues applied the stock model to show that the cumulative release of a trillion tonnes of carbon[12] would equate to a rise in 2°C in the surface temperature of the planet, but as a median within a broad distribution of outcomes, both higher and lower. This work has been subsequently included in the 2014 Intergovernmental Panel on Climate Change (IPCC) 5th Assessment Report (Working Group I, Climate Science) which introduced it into the mainstream climate debate.

As long as the total carbon released remains less than a fixed amount over, say, a 500- to 1000-year period, the rise in temperature is contained, at least to a level that is equivalent to that amount.

Imagine, if you can, the atmosphere's cumulative carbon emissions, the driver for surface temperature change, as a 1 trillion tonne container, or a jug, being filled. As you keep pouring, the container fills and a second container is needed. You've just passed 2°C of warming. In reality, that point may be passed before 2040 and by the end of this century be well on the way to filling the second container. The world has the fossil resource to do this, and most of the necessary infrastructure already in place to extract and use it.

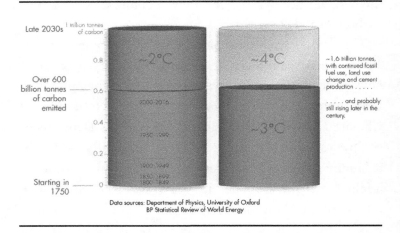

Data sources: Department of Physics, University of Oxford
BP Statistical Review of World Energy

The level in a container can also be reduced, in this case by removing carbon from the atmosphere. Nature does this slowly over time, but this could be accelerated by human intervention. Direct air capture of carbon dioxide is technically possible and when combined with geological storage of that carbon dioxide, the process is reversed, a bit like draining a bathtub as the water continues to run in. Depending on the relative rates of addition and removal, this may slow the accumulation, level it out or even see it drop. Today, the process to capture carbon dioxide directly from the air is in its infancy, although the capture and geological storage of carbon dioxide from industrial sources is operating in many locations around the world.

The release by the start of 2017 of nearly 610 billion tonnes of carbon has committed society to a certain level of warming already. But seeing this warming as a consistent signal in the global temperature data can be difficult; sometimes the trend can remain hidden.

More immediate data noise in the temperature record can be created by shorter climate events, such as the El Niño Southern Oscillation (ENSO). This is a periodic upwelling of warmer (El Niño event) and cooler (La Niña event) water in the Pacific that is on such a scale it slightly shifts the global temperature. In 1998 the world experienced what is categorised as a 'very strong El Niño' and global temperatures peaked as a result. As the 2000s and 2010s unfolded, the US National Oceanic and Atmospheric Administration (NOAA) reported only one modest El Niño year, but seven La Niña years, which meant that the temperature record showed little sign of an overall warming trend. This in turn led to considerable criticism of climate science and may have even partly contributed to the collapse of the Copenhagen Climate Conference in 2009.

But everything changed in 2015 with the emergence of another 'very strong El Niño' event. In May 2016 *The Economist* featured an article on the El Niño in the Pacific and the impact or otherwise that a warming climate system might be having on it. The author noted that the recent sweltering temperatures may help settle debates over a supposed 'pause' in global warming that occurred between 1998 and 2013.

During that period the Earth's surface temperature had appeared to rise at a rate of only 0.04°C a decade, rather than the clear 0.18°C increase of the 1990s, but in reality there had been no pause at all.

Global temperature data can be challenging to analyse, but a very simple analysis I put together shows quite a clear result. As already noted, El Niño events can be categorised, with the events of 1997–1998 and 2015–2016 both listed as Very Strong. 1972–1973 and 1982–1983 were also Very Strong events, giving a total of four such events over the last 40 years. Each of these events led to a temperature spike in the global record as reported by NOAA.

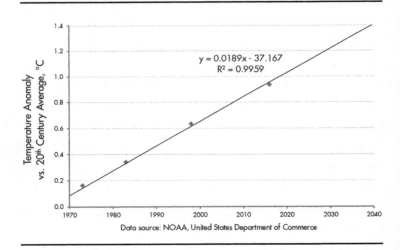

A plot of this data shows an almost perfect linear trend over recent decades, with the Very Strong El Niño same year global temperature anomaly rising monotonically at 0.19°C per decade. There is no sign of a pause in warming or acceleration, at least over the last 40+ years. Extending the trend into the 2030s indicates that a future Very Strong El Niño event in that period would result in a 1.3°C temperature rise against the 20th-century average, which is about the equivalent to 1.5°C above pre-industrial levels in the NOAA time series (using late 19th century as a proxy for pre-industrial).

Casting back a bit further to a much earlier Very Strong El Niño, brings us to 1926. This was reportedly an extreme event for the period and corresponded with the most severe drought in tropical South America during the 20th century. Including it in the chart as well as further points in 1958 and 1966 (both Strong events), shows the current linear trend still holding back to 1960s, but not into the 1920s. At this time atmospheric carbon dioxide was only just beginning to rise. But the fact that the 1920s El Niño is matched by a presumably elevated 1960s El Niño perhaps points to just how severe that event must have been.

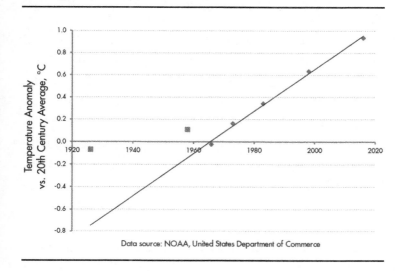

Data source: NOAA, United States Department of Commerce

The data for the last half century for comparable El Niño event years and coinciding with a rise in atmospheric carbon dioxide from 315 to 400 ppm (vs. 275 to 315 ppm over the century before that) indicates that the underlying surface temperature trend is rising consistently, despite the noise associated with year on year fluctuations of the Southern Oscillation (El Niño and La Niña) and other phenomena. This data noise has given rise to claims of both no global warming and accelerating global warming. The reality is sobering enough, even without the histrionics from some observers.

But a further question to ask is when the current upward trend really started in earnest and therefore what is the temperature rise today against a pre or early industrial (i.e. 1750−1850) baseline. This is an important question as there is now the established desire to keep warming well below 2°C, but with the real prize being to limit this even further and ideally to 1.5°C.

If the current strong warming trend started in the middle of the last century, say in the post-war boom, then at 0.18°C per decade this results in warming over the period of nearly 1.2°C. The 1950s were also presumably warmer than the 1850s as carbon dioxide levels had risen by some 30 ppm over that period, which argues for the current level of warming to be something more than 1.2°C.

Another way of looking at this is to return to the relationship between cumulative carbon and temperature. With a bit over 600 billion tonnes of carbon emitted since 1750, the Oxford University model would imply warming of about 1.2°C.

Yet another way is to seek an answer from a group of climate scientists. At the 39th Forum of the MITJP I posed the question and one respondent (the Chatham House Rule applies) argued that current warming is around 1.1°C since pre-industrial times, but that there is more to the story than this. The climate system is not at equilibrium, with the oceans still lagging in terms of heat uptake. Therefore, if the level of carbon dioxide in the atmosphere was maintained at some 400 ppm, the surface temperature would rise by another few tenths of a degree before the system reached an equilibrium plateau. That would take us perilously close, if not over, the 1.5°C goal of the Paris Agreement. This implies that 1.5°C is only possible if there is a fall in atmospheric carbon dioxide, back below 400 ppm. But note that carbon dioxide is currently rising at 2−3 ppm per annum.

Thinking about climate change as a stock problem rather than a flow problem changes the nature of the solution and the approach. Although emissions in 2020 or 2050 may be useful markers of progress, they do not guarantee success, as they are measures of flow, not stock. For example, meeting a 2050 global goal of reducing annual emissions by 50% relative to 1990 would be a remarkable achievement. This wouldn't, however, result in a change in the eventual level of warming if the same cumulative amount of carbon was subsequently released, albeit at a slower rate than might otherwise have been the case. Looking at the issue through a stock lens also changes the view of mitigation opportunities.

The stock perspective means that a truly comprehensive international treaty really needs to consider the point at which carbon dioxide emissions reach zero, or net-zero, being a zero sum of emission sources and sinks. When net-zero emissions is reached, accumulation ceases and the level of carbon dioxide in the atmosphere stabilises. Remarkably, the Paris Agreement did incorporate such a goal. It didn't, however, consider how to remove some of the carbon from the atmosphere; that may be a task for the next generation of negotiators.

The climate issue doesn't begin and end with carbon dioxide; it is one of many greenhouse gases. The most prevalent is water vapour, but it changes dynamically with temperature and so reinforces the role that carbon dioxide plays. But several other greenhouse gases have become problematic over the last century as a result of human activity. Perhaps the most important of these is methane. Anthropogenic emissions are about 250 million tonnes per annum, a small fraction of that for carbon dioxide. The gas is naturally emitted from rotting vegetation, some animals and natural vents in the Earth's crust, but human activities have added to this. The largest anthropogenic sources are agricultural methane from belching cows as they digest food, rice paddies and fossil methane (natural gas) from fugitive sources in the global oil, gas and coal industry.

Methane is important because it is a more potent greenhouse gas than carbon dioxide. But methane does not accumulate in the atmosphere like carbon dioxide; rather it breaks down quite quickly with a 'half-life' of about seven years, so on a 100-year basis (with the methane effectively gone and instead existing as carbon dioxide) the warming impact of a tonne of methane emitted now compared to a tonne of carbon dioxide is about 28 times. When the same calculation is done on a 20-year basis that factor rises to about 80, because methane emitted now has an immediate impact on atmospheric warming that is over one hundred times greater than a tonne of carbon dioxide.

While agricultural methane may require real lifestyle changes to bring down, e.g. less meat and rice consumption, industrial methane emissions management can be implemented. Often mitigation may be a case of good housekeeping, such as monitoring and maintaining pipelines to minimise small leaks.

Methane is different from carbon dioxide because it presents us with more of a flow problem than a stock problem. The points in time at which methane and carbon dioxide are emitted and the shape of any reduction pathway relative to each other are important. It also means that bundling methane and carbon dioxide together may lead to poorly designed policy frameworks to manage them.

Peak warming is largely dictated by the cumulative amount of carbon dioxide emitted over time. If a certain amount of methane is also emitted, the timing of that emission is what matters. Methane that is emitted today will immediately impact the rate of warming, but long before peak warming is reached (assuming carbon dioxide emissions are eventually brought under control and warming actually peaks) the methane will have left the atmosphere and been converted to carbon dioxide, in which case its impact on peak warming is based only on the carbon dioxide that remains from the methane. Warming may have accelerated in the short term, but peak warming will remain largely unchanged. In this case, the warming potential of methane, expressed in terms of its impact on peak temperature falls sharply and comes close to the stoichiometric conversion of methane to carbon dioxide, which is about 3 — that is, a tonne of methane when combusted or oxidised in the atmosphere gives rise to about three tonnes of carbon dioxide. Conversely, methane that is emitted much later, say when peak warming is much closer, will directly add to whatever level of temperature might have been reached as a result of carbon dioxide acting alone.

Does this mean that society shouldn't bother about methane today? Unfortunately, the answer is an ambiguous one. If there is confidence that society will quickly and decisively reduce carbon dioxide emissions, then of course methane and other greenhouse gases must be reduced sharply in the near term as well. If society doesn't act on these gases, then that problem will still be present at the time peak carbon dioxide-induced warming occurs, in which case peak warming will almost certainly overshoot the goal of 2°C, with the additional warming from the other greenhouse gases. But if the carbon dioxide issue isn't addressed, then addressing the methane issue now doesn't offer a lot of benefit for later on. Instead, the benefit is less short-term warming, as the more intense burst of heat that the methane is providing will have been removed.

Of course, since it isn't known how well or otherwise the task of carbon dioxide mitigation will proceed (despite the fact that the track record is pretty poor), there is still a feeling of obligation to act on methane now in case carbon dioxide mitigation picks up. At least the near-term rate of warming should slow slightly. However, treating methane as if it were interchangeable with carbon dioxide but with a convenient and high multiplier to make us feel that modest effort is delivering great benefit, could hide the reality that little progress towards limiting peak warming is being delivered at all.

A story with similar claims of great progress emerged in October 2016 with the agreement of an addendum to the Montreal Protocol (agreed in 1987 to progressively eliminate the use of CFCs, coming into force in 1989), which will bring the hydro-fluorocarbon (HFC) family of gases into that process, leading to their eventual elimination from day-to-day use.

HFCs have been increasingly used this century as an alternative to ozone-damaging CFCs in refrigeration systems. Though HFCs provide an effective alternative to CFCs, they are also powerful greenhouse gases. A snapshot of current greenhouse gas emissions to atmosphere highlights the HFC issue; today they represent approximately 1billion tonnes on a carbon dioxide equivalent basis, or about 2% of the total climate problem.

While 2% is not insignificant, being just above the impact that aviation currently has on the climate issue, it is the longer-term impact that the growth in use of these products has that is the main cause for concern. India alone could build upwards of 400 million refrigerators over the coming 20 years. But following seven years of negotiations, the 197 Parties to the Montreal Protocol reached a compromise under which developed countries will start to phase down HFCs by 2019. Developing countries will follow with a partial freeze of HFCs consumption levels in 2024, with some countries freezing consumption in 2028. By the late 2040s, all countries are expected to consume no more than 15−20% of their respective baselines. A small group of countries is treated more leniently owing to the very high local temperatures experienced during much of the year.

This is an important agreement, but the claim that came from the United Nations Environmental Programme (UNEP) that the Kigali Amendment to the Montreal

Protocol (as this was agreed at the 28th Meeting of the Parties to the Montreal Protocol at their meeting in Kigali, Rwanda) was the equivalent of saving 0.5°C from the anticipated warming of the climate system was something of a surprise. Secretary of State John Kerry rounded this up slightly and referred to the result as a 1°F achievement.

The source of the 0.5°C figure is a series of academic papers[13] that note that most HFCs now in use have relatively short lifetimes in the atmosphere in comparison with carbon dioxide — so, like methane, they are also short-lived climate pollutants (SLCP). The global average lifetime, weighted by the production of the various HFCs now in commercial use, is about 15 years, with a range of 1 to 50 years. But the chemicals have very high global warming potentials (e.g. over 4000 for HFC-143a).

Therefore, HFCs have a significant impact on warming in the years immediately after release into the atmosphere, but their impact on peak warming, whenever that occurs, depends more on the level of the gas in the atmosphere at that time rather than the emissions now or in the near term. Like methane, the warming that HFCs induce is nearer term and tends to mean that a certain temperature is reached earlier than might have been the case, even if the eventual peak warming remains the same.

Nevertheless, some significant warming numbers arise, although they are very dependent on the HFC demand projections and scenarios that are developed. Most importantly though and as is the case for methane, HFC mitigation should not be viewed as an alternative strategy for avoiding the 2°C peak warming, but rather as a critical component of a strategy that is heavily dependent on mitigation of carbon dioxide. While the claims made after the meeting in Kigali were not wrong, they nevertheless mixed stock and flow impacts and called them one.

CLIMATE IMPACTS

We now know that a rise in the level of carbon dioxide and other greenhouse gases in the atmosphere will cause the surface temperature to rise, but then what?

The average surface temperature is one of the principal climate factors that influence the shorter-term weather patterns that affect our daily lives and well-being. The surface temperature sets the amount of water that exists as ice at the poles and as a gas in the atmosphere. Recently though, perhaps in an effort to raise the profile of the climate issue during the period up to 2015 when there wasn't an apparent upward trend in the surface temperature, it's interesting to note that almost every extreme weather event is blamed on climate change.

Through 2010 and 2011 in particular, weather extremes came to dominate the headlines. Extreme drought, rainfall, flood and wind all played a role in making the period one of the most expensive in terms of damage to infrastructure. In some locations there was significant loss of life; at least 35 people died in the floods in Australia. This period also saw the subject of extreme weather events rise up the climate change agenda, with numerous academic papers, blogs, seminars and campaigns focused on the issue.

Since then we have continued to see cyclones, floods, extreme events, drought and heat waves. But are they all indicative of a changing climate, as some would have us believe? Unfortunately, the answer is yes and no.

Certainly as the atmosphere gains energy (i.e. as it warms), changes in weather patterns can be expected. Warming is altering the jet stream and changing atmosphere and ocean temperature differentials that lead to storms. A warmer atmosphere will hold more moisture, which can mean higher levels of precipitation when it rains.

Consequently, we can expect to see an increase in weather events that fall outside accustomed weather patterns. The problem here is that there have always been extreme events and there have also been periods of bunched extreme events. This may be driven by climate cycles, such as ENSO. Returning to the bunching of extreme events in 2010 and 2011, when there was a strong La Niña event (part of the ENSO cycle), a similar bunching can be found in 1974–1975 when, again, there was a strong La Niña.

For argument's sake, let us compare the 1974 and 2010 timelines in which there are similar cases of intense flooding in Queensland Australia, super tornado activity in the United States and drought activity in East Africa. Similarities exist, although the severe droughts that affected the south-western US states didn't occur in the 1970s. Rather, the 2011 Texas drought has been shown as exceptional by any standard. Extreme weather events deserve our attention, but there needs to be increased diligence when it comes to directly associating them with climate change.

The variability in global temperatures (seasons, weather) can be approximated as a normal (Gaussian) distribution, the so-called bell curve.

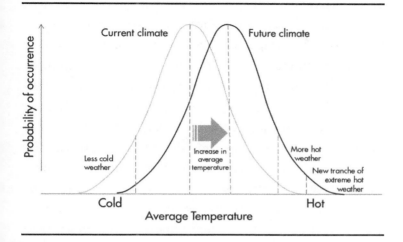

A normal distribution has 68% of the observations falling within one standard deviation[14] (σ) of the mean value. The tails of the normal distribution decrease quite rapidly so that 95% of the observations fall within two standard deviations of the mean. About 99.9% fall within three standard deviations of the mean. For example, if the summer average maximum temperature in a given region of the world is 25°C with a standard deviation of 4°C, then there is an expectation of only one day every 10 years (0.1% based on 100 day summers) with a maximum temperature above 37°C (25 + 3 × 4).

In a warmer climate where the average in the example is now 29°C, statistically there will be five days every year where the maximum temperature exceeds 37°C and one day every 10 years where it exceeds 41°C, assuming that the shape of the distribution remains the same as it shifts to the right.

Following the extreme European heat wave of 2003, Professor Myles Allen and his team at Oxford University produced an analysis that showed that the event lay well beyond the normal two standard-deviation band around the historical average. Therefore, it was unlikely to have occurred without a certain level of background warming.

A similar analysis of temperature data over the period 1951–2011 by NASA climatologist James Hansen explores the phenomenon in considerable depth and shows that extreme heat events should be a cause for concern. Hansen has shown that the distribution of seasonal temperature has shifted, leading to an increase in extreme events. An important change is the emergence of a category of extremely hot summertime outliers, more than three standard deviations warmer than the average 1951–1980 baseline. This extreme, which covered much less than 1% of Earth's surface in the base period, now typically covers about 10% of the land area. Hansen concludes that extreme heat waves, such as that in Texas and Oklahoma in 2011 and Moscow in 2010, were attributable to global warming, because their likelihood was negligible prior to the recent rapid global warming.

The increase, by more than a factor 10, of area covered by extreme hot events in summer reflects the shift of the distribution in the past 30 years of global warming. Hansen has concluded that the extreme hot tail of the temperature distribution has shifted to the right by more than one standard deviation in response to warming over the past three decades. He goes on to say that additional global warming in the next 50 years would see the one-in-ten-year event becoming the norm, with what were one-in-a-century events becoming quite common.

The likelihood of a current summer being 'hot' by comparison to the 1960s is now 80%. This change is sufficiently large that anyone old enough to remember the climate of the sixties, or part of it, should recognise the existence of climate change. Others may be starting to recognise changes in weather patterns, but it remains debatable how long it will take the rest of us to catch up. An increased incidence of what were once-in-a-century events may well trigger such a reaction, although such a change may not be apparent until the 2020s or 2030s. Although the hydrological cycle will undoubtedly change as a result of rising temperatures, there is still much research to be done to understand what this means in terms of global, regional and local precipitation.

As carbon dioxide levels in the atmosphere continue to rise, there is a great deal said about storms and droughts but not as much about rising sea levels, although the more recent emphasis on this by the Alliance of Small Island States (AOSIS) in the lead-up to COP21 certainly made the news. Miami is also raising the profile of sea level rise as it responds to ocean encroachment in certain parts of the city with a $400 million capital works programme.

In fact, a great deal is known about sea level, carbon dioxide and temperature, thanks to the paleoclimate record. This was the subject of the first session of the 31st MIT Global Change Forum in 2010 in association with Université Catholique de Louvain. The focus was on the interglacial periods of the last million years, which saw global temperatures and carbon dioxide at levels similar to the Holocene period of the last ten thousand years. These relatively warm periods occur every 100–150,000 years (we are in one now) as a result of the additional heat falling on the planet from small variations in the Earth's orbit around the sun.

The most recent interglacial peak was the Eemian, about 130,000 years ago, with a temperature peak at the top end of the interglacial range or about 2°C above pre-industrial Holocene levels, but with the carbon dioxide level around 300 ppm. There is good evidence[15] that the sea level in this period topped out at some 6−10 metres above current levels, before plunging well over 100 metres as glaciation took hold in the northern hemisphere. This raises the issue of where sea levels might peak as a result of the current elevated levels of carbon dioxide and related warming. The analysis presented at the Forum was also about where so much water would come from.

Certainly, there is enough ice in Greenland to raise sea level another six metres, but Greenland didn't melt completely during the Eemian period as scientists have drilled and extracted much older Greenland ice cores. Rather, the evidence points to a partial melting of the Greenland ice sheet followed by the collapse of the West Antarctic Ice Sheet. The former contributed some 2.2−3.4 metres of sea level rise and the latter 4.4−6 metres. During this period ice core measurement indicates that the Polar Regions of the northern hemisphere experienced temperature rises of 3−5°C, in response to a global temperature excursion of about 2°C. The rate of sea level rise was 5−6 mm/year or ~5.6 metres over 1000 years. This sounds like an extraordinary amount of time, but we shouldn't forget that the additional warming, driving the change today, is likely to be considerably more than during the Eemian period.

Estimates of sea level rise vary significantly, with some studies showing an overall change of up to 2 metres[16] by the end of this century. Such a rise could be very disruptive and require considerable relocation of people and infrastructure.

Sea level won't rise by more than 85 metres; it is limited by the amount of water stored in Antarctica, Greenland and the remaining non-Arctic glaciers. While there is no suggestion that all of this will melt, temperature rises of 2°C, 3°C or 4 + °C could see sea level rising by tens of metres over the coming centuries. Under such circumstances, many cities might even have to migrate away from their coastal locations as captured so vividly in the film adaptation of David Mitchell's century-hopping novel, *Cloud Atlas*. There is a scene set in a futuristic Neo-Seoul in which the characters are looking out over the city onto what appear to be partially submerged skyscrapers in the distance. One of the characters makes the comment that it is Old Seoul and that if the seas keep rising as they have been, Neo-Seoul itself will be under water in a century.

Should this amount of sea level rise eventuate, one can only speculate on the types of mega-projects that may materialise in the centuries to come. Perhaps a future generation will have to dam the Straits of Gibraltar across the shallowest part of the straits to protect the coastline of the Mediterranean and Black Sea.

3

COUNTING CARBON

POTENTIAL PATHWAYS

The very large-scale anthropogenic release of carbon dioxide to the atmosphere has been increasing for over 250 years. The question we collectively face is, how much can be emitted by society before a rise in the surface temperature triggers changes in sea level and the climate that become problematic, or worse, to adapt to?

Scientists and politicians have been considering this question for some time. Not surprisingly it sits at the heart of the international process on climate change managed through the UNFCCC. At the COP in Cancun in 2010, delegates agreed on text that called for deep cuts in global greenhouse gas emissions, with a view to reducing emissions so as to hold the increase in global average temperature below 2°C above pre-industrial levels. The Paris Agreement has strengthened this even further, with an explicit goal to limit warming to well below 2°C.

So what exactly is 'well below'? It could mean 10–25% below the number mentioned. This means that the Paris Agreement seeks a temperature in the range of

1.5−1.8°C, rather than 2°C. The Agreement has also embraced the concept of 'net-zero emissions', describing this as the need to achieve a balance between anthropogenic emissions and sinks of greenhouse gases.

As that final gavel came down in Paris, cumulative emissions since the beginning of the Industrial Revolution amounted to nearly 600 billion tonnes of carbon.[17] Using the previously defined relationship developed by Allen et al. and starting with 600 billion tonnes, a 1.5°C temperature rise would potentially be exceeded if cumulative carbon emissions passed 750 billion tonnes, which could happen as early as 2028 if emissions are maintained at current levels.

Yet the 2°C reference isn't as clear-cut as it appears. The Paris Agreement may be explicit in citing it, but turning this into something tangible in terms of global emissions and national objectives is difficult. It requires an understanding of complex statistics. This was highlighted by another paper from Professor Allan and his colleagues.[18] They showed that the uncertainty of the climate response, combined with a variety of emission pathways, allows only probabilities for staying below 2°C to be determined, rather than a precise outcome.

This is a useful analysis, but how should it be applied to the global goal? The call in Paris was to hold the increase to well below 2°C, but this means different things to different people, with some interpreting it as a 'reasonable probability' which might then be interpreted as a 75% chance or better. This implies an emissions limit between the years 2000 and 2049 of less than 1000 Gt of carbon dioxide. But with carbon dioxide emissions from 2000 to 2016 already totalling about 600 billion tonnes or over half that amount, it is difficult to view 'well below 2°C' with a high probability outcome as a realistic objective. The amount would be reached with just 10 more years of emissions at 2016 rates — that is, as early as 2027. In other words, staying within this limit supposes a trajectory that sees no further rise in emissions after 2016, but instead year-on-year reductions of nearly 2 billion tonnes until emissions are near zero. Delaying the peak until 2020 pushes up the reduction rate to 3 billion tonnes per annum.

By contrast, accepting a 50% chance of staying below 2°C gives a very different outcome. If emissions peak in 2020, a reduction pace of 1 billion tonnes per annum is then required. Alternatively, should emissions plateau in 2020 and start reducing in 2025, the annual effort rises to 1.5 billion tonnes. These are still very challenging numbers, but at least more plausible than the 75% probability case. Given where the world is today, achieving the 75% case is highly unlikely.

In my experience, any two people who talk about 2°C have very different perspectives on likelihood, usually without any thought as to the implications behind their assumptions. The EU Climate Change Expert Group is clearer on this in its representation of the 2°C goal, where it specifically notes the 50% probability case in its key messages.[19]

The 2°C goal stems from an integrated assessment of climate data based on a multitude of factors. It appears to be the point at which various systems may see a step change in their response to rising temperatures. This includes the collapse of ice shelves, changes in major ocean current circulation, the demise of marine ecosystems and extensive coral bleaching. Much of this was summarised in a document from a 2005 conference, *Avoiding Dangerous Climate Change*, convened by the British Government prior to their hosting of the G8 that same year. Although the EU had proposed a 2°C target prior to 2005, it was at this conference and the following G8 meeting where it gained political traction, leading to Cancun in 2010 where it was agreed as a formal goal, since the objective of the UN Framework Convention is to avoid dangerous anthropogenic interference with the climate system.

There is no guarantee that society can collectively limit the temperature rise to well below 2°C. Even if emissions stopped today, the range of possible outcomes from the current 400 ppm carbon dioxide level includes the possibility of exceeding 2°C, albeit at a very low level (less than 10%, so perhaps not that low). This is because the atmosphere is not in a state of heat equilibrium and will continue to warm after a given level of carbon dioxide is reached. As such, determining a target atmospheric concentration of carbon dioxide is difficult. 450 ppm, a level often equated with 2°C, is convenient in that it is above current levels, was feasible (at a stretch) when first raised and equates to a 50% chance of limiting the temperature rise to 2°C. More recently NASA's James Hansen[20] has argued for a goal of 350 ppm, to restore the current heat imbalance in the earth system and stop a rise in temperature. The problem with this goal is that it has already been passed and nobody likes setting unrealistic objectives.

Unfortunately, the use of a 2°C reference point has now become politically rigid, to the extent that it is hardly possible to explore what the number really means, let alone talk of surpassing it. The most common argument is that if 2°C is passed, then society must be on a certain pathway to pass the 4°C mark as well — although this would require a century of complacency. This line of thinking is highlighted by, amongst others, the World Bank in a 2012 report,[21] where the reader is presented with the stark choice of a 2°C or 4°C world by 2100. In their 5th Assessment Report launched in 2014, the Intergovernmental Panel on Climate Change (IPCC) also chose to present these two scenarios in order to contrast decisive action with a failure to act. Arguably, though, the space between 2°C and 4°C may turn out to be where the real story lies.

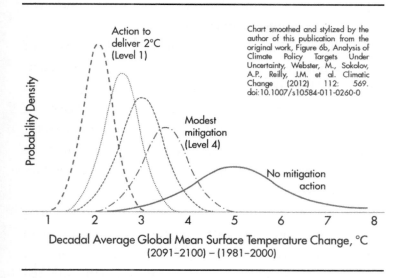

Action to deliver 2°C (Level 1)

Chart smoothed and stylized by the author of this publication from the original work, Figure 6b, Analysis of Climate Policy Targets Under Uncertainty, Webster, M., Sokolov, A.P., Reilly, J.M. et al. Climatic Change (2012) 112: 569. doi:10.1007/s10584-011-0260-0

Probability Density

Modest mitigation (Level 4)

No mitigation action

Decadal Average Global Mean Surface Temperature Change, °C (2091–2100) – (1981–2000)

That space was comprehensively analysed in 2009 by the MITJP.[22] The authors demonstrated that even a modest attempt to mitigate emissions could profoundly affect the risk profile for equilibrium surface temperature. They looked at five mitigation scenarios, from a *'do nothing'* approach to a very stringent climate regime (Level 1, akin to a 2°C case). In the 'do nothing' approach, mid-range warming by the end of the century is some 5°C compared to the late 20th century, but with a wide distribution, which means that there is a small probability of warming up to 8°C or more — an unacceptably high outcome even when accounting for the small probability that it might occur. But even modest mitigation efforts, while not shifting the mid-range sufficiently for an outcome close to 2°C, nevertheless radically changes the shape of the distribution curve such that the spread narrows considerably, with the highest impact outcome dropping by some 5°C. As mitigation effort increases and the mid-range approaches 2°C, the distribution narrows further such that the highest possible outcome is limited to 3°C.

This is not an argument for limiting the global effort to modest mitigation, but recognition that if modest mitigation is the best that can ultimately be achieved, the risk reduction it delivers has very high value to society. It highlights that a singular focus on a very difficult-to-achieve goal can be counterproductive if it results in complete breakdown of the efforts and therefore a failure to achieve even a less ambitious goal.

The concept of a 2°C threshold has been given new meaning in recent years with the argument that any fossil fuel use that takes society beyond such a level should not take place. However, the debating position being adopted usually avoids discussion of the energy transition necessary to ensure this use doesn't need to happen. Rather, it is formulated as a simple call to leave the fossil fuels in the ground such that the 2°C container doesn't overflow into a second container and commence the journey towards 4°C.

An alternative way of looking at this issue is to consider the so-called lock-in layers within the 2°C container. Each layer represents a chunk of infrastructure in use today that is likely to continue operating until the end of its normal life, emitting carbon dioxide and therefore adding to the growing accumulation of carbon dioxide in the atmosphere. Major layers are described below:

1. The largest existing commitment is coal-fired power stations. There is some 2000 GW of coal-fired capacity in existence (in 2016), with each GW emitting about 6 million tonnes of carbon dioxide per annum. More than half of this has been built in this century, so we might assume an average age of 20 years for the existing facilities. That leaves about 30–40 more years of operation. Even assuming that no more are built, that means cumulative carbon dioxide emissions of over 400 billion tonnes from 2016 onwards, or 115 billion tonnes of carbon. But utilities could well build another 500 GW in rapidly developing countries such as India and South Africa, which would add another 150 billion tonnes of carbon dioxide, or 40 billion tonnes of carbon.

2. There are about 1 billion passenger cars in the world today and production in 2015 was approaching 70 million per annum. Assuming the average age of a current world car is 7–8 years and the average lifetime of a car is 15 years, this population could emit a further 10 billion tonnes of carbon. Even if internal combustion vehicle production ended in 20 years, another billion vehicles will almost certainly be built in the interim, which, in turn, could add a further 16 billion tonnes of carbon to the atmosphere as they are used.

3. Natural gas use in power generation is growing rapidly, with some 1600 GW in use in 2010, growing to 2000 GW over the 2010—2020 decade. Even though a gas-fired power station emits less than half the amount of carbon dioxide as a similar sized coal plant, this fleet could see a further 150—200 billion tonnes of carbon dioxide or nearly 50 billion tonnes of carbon emitted prior to retirement, in addition to the 115 billion tonnes of carbon from coal listed above.

4. According to the IEA (CO_2 Emissions from Fuel Combustion 2016), residential use of gas results in almost 1 billion tonnes of carbon dioxide emissions per annum. This is somewhat hard-wired into cities, so it is difficult to dislodge any time soon (although having replaced our gas boiler at home with an electric one because of new UK flue regulations, I know that it's clearly not that difficult for an individual home to do so). Nevertheless, this could well continue for 30—40 years, so adding perhaps another 10 billion tonnes of carbon.

5. Aviation and shipping not only have an existing fleet but also show almost no signs of finding viable large-scale routes to zero emissions. Biofuels may be the solution for both. Expect another 40 years of emissions at a minimum, which is another 10 + billion tonnes of carbon.

6. Finally, there is the manufacturing industry, which emits 6 billion tonnes of carbon dioxide per annum globally. This includes refineries, ferrous and nonferrous metal production, cement plants, chemical plants,

the pulp and paper industry and various other sectors. Capacity is renewing rapidly not only because of growth and development but also because of the gradual decline of developed country capacity in favour of much larger and more efficient production in regions such as the Middle East. New capacity will operate for 30 to 50 years at least, so this sector could be responsible for another 120 billion tonnes or more of carbon dioxide or about 32 billion tonnes of carbon.

The sum of these 'emission lock-in layers' adds up to nearly 300 billion tonnes of carbon, which gives future minimum cumulative emissions of some 900 billion tonnes. This is in excess of a notional 1.5°C carbon emissions threshold and is knocking on the door of 2°C, or a trillion tonnes of carbon. The calculation includes the major sources of emissions (e.g. small oil-fired power stations are not included) and probably represents the earliest case for retirement of existing power stations and industrial facilities. Many could just keep running for decades longer than expected. Staying within the trillion-tonne threshold, therefore, leaves little room for complacency regarding the next generation of assets and other emissions sources, such as land-use change.

We have all watched China grow rapidly on the back of large-scale manufacturing while building vast swathes of infrastructure from cities such as Shanghai and Chongqing to the high-speed rail networks that now connect them. Between 1995 and 2015 cumulative emissions from China amounted to some 130 billion tonnes of carbon dioxide, or 100 tonnes per person. For the most part, this wasn't for personal domestic use (i.e. home electricity and heating), but to make products for consumers at home and for export, which in turn has financed domestic infrastructure. Yet now China is starting to look to other economies to make its raw materials and supply finished products as it attempts to develop its service sector.

The situation in those less developed economies is not dissimilar to China 30 years ago. Some 3 billion or more people live with little or modest levels of infrastructure. While they may have rudimentary renewable energy for lighting and other services, their standard of living remains far below other parts of the world. Their development pathway may well be similar to that of the Chinese from the 1980s onwards. Products made for the Chinese economy as the Chinese service sector grows and energy use reaches a plateau might even fund that pathway.

The 100 tonnes per person of emissions related to development includes the hardest to decarbonise. These emissions come from steel mills, cement plants, chemical plants, manufacturing industry and heavy goods transport. These are the backbone industries and services for development, many of which have disappeared from developed economies. They are also expensive to decarbonise, which is problematic for economies in the earlier stages of rapid development. This development also leads to a degree of lock-in because once industries are created and jobs are in place, governments often go to great lengths to retain them. The same industries are also needed to make a wide range of products, from cars to iPhones, for consumers in the global market.

One particular consideration for the implementation of the Paris Agreement is the 100 tonnes per person of development emissions and the lock-in that follows. If this is applied to 3 billion people, it could result in another wedge of emissions approaching some 100 billion tonnes of carbon. Adding this to the 900 billion tonnes calculated above quickly takes the total above 1 trillion tonnes.

While the net-zero goal looks feasible and can be imagined as a longer-term outcome, there may be an interim emissions bulge as development continues and the supporting industries required for infrastructure are put in place. This unfortunately may take society beyond 2°C rather than leave it well below.

In the months before negotiators convened in Paris, a flood of national proposals for managing and reducing emissions arrived at the UNFCCC offices in Bonn. These were the Nationally Determined Contributions (NDC, but initially prefixed with 'I' for 'intended'), in effect the principal mechanism of the Paris Agreement. The number of submissions increased to include nearly every country by the early days of December, although at the time of the cut-off point for the pre-Paris assessment of progress promised by the UNFCCC Secretariat, some 150 had been received. That represented well over 80% of global emissions.

The stark reality of these submissions is that while laudable and ambitious in intent, not a single NDC reflects a pathway that is now aligned with the ambition of the Paris Agreement. Some, assuming full execution, come close to a pathway that might be considered close to 2°C, but Paris subsequently shifted the goalposts to around 1.5–1.8°C of warming compared to pre-industrial levels.

In their 30 October 2015 press release outlining the results of their NDC assessment, the UNFCCC announced that the NDCs have the capability of limiting the forecast temperature rise to around 2.7°C by 2100, which they noted was much lower than the estimated four, five or more degrees of warming projected by many prior to the NDCs.

This was quite a surprising outcome and the finding by the UNFCCC added to the Paris momentum. The number of contributions submitted was astonishing as was the belief that if a shift from 4 +°C to 2.7°C could be achieved in one attempt at a national level, then surely something profound could result from a more ambitious attempt.

There is no doubt that the combined impact of the 190 or so NDCs, if and when fully implemented, will represent a departure from a business-as-usual emissions profile. The full UNFCCC NDC report goes into considerable detail to show this, concluding that global emissions in 2025 could be 2.8 Gt carbon dioxide equivalent (accounting for other greenhouse gases such as methane, but in equivalent terms) less than a pre-NDC pathway and 3.6 Gt in 2030. But the range associated with each projection is significant, with the 2030 number ranging from no change to as much as 7.5 Gt. The projected total global emissions for 2030 would still be 56.7 billion tonnes of carbon dioxide equivalent. Notably, the full report makes no mention of the temperature impact that this would have. Rather, the report summary only noted that the NDCs contribute to lowering the expected temperature rise until and beyond 2100.

So how did the UNFCCC arrive at a temperature projection of 2.7°C? What approach might they have used to add up the cumulative carbon? It is very clear from their report that the authors were not prepared to do the required calculation. They even state that making the necessary assumptions for such an outlook is beyond the scope of the report.

Assessing the NDCs in aggregate requires some thinking about methodology. For starters, the temperature rise that will be seen in 2100 is driven by cumulative emissions over time (with a climate sensitivity of about 2°C per trillion tonnes of carbon — or 3.7 trillion tonnes carbon dioxide), not emissions in the period from 2020 to 2025 or 2030 which is the point at which most of the NDCs end. In fact, 2025 or 2030 represents more of a starting point than an end point for many countries. Nevertheless, in reading the NDCs, the proposals put forward by many countries give some clues as to where they might be going.

For Europe, the United States and a host of other developed economies, the decline in emissions is already underway or at least getting started, with most having stated that by mid-century, reductions of 70−80% versus the early part of the century should be possible. Many emerging economies are also giving signs as to their long-term intentions, but the information supplied is sketchy. For example, the South Africa NDC proposed a Peak-Plateau-Decline strategy, which sees a peak around 2020−2025, then a plateau for a decade, followed by a decline. China has signalled a peak in emissions around 2030 but has not offered a decline pathway after that date. With development at a very different stage, the Indian government has yet to communicate a peak date.

Nevertheless, thanks to some bold and perhaps optimistic assumptions, it is possible to assess the cumulative efforts and see where the world might be by the end of the century. In doing this, I have employed the following methodology:

- Use an 80/20 approach, i.e. assess the NDCs of the top 15–20 emitters and make an assumption about the rest of the world. My list includes the United States, China, India, Europe, Brazil, Indonesia, South Africa, Canada, Mexico, Russia, Japan, Australia, Korea, Thailand, Taiwan, Iran and Saudi Arabia. In current terms, this represents 85% of the global energy system carbon dioxide emissions.

- For the rest of the world (ROW), assume that emissions double by 2040 and plateau, before declining slowly throughout the second half of the century.

- For most countries, assume that emissions are near zero by 2100, with global energy emissions nearing 5 billion tonnes. The majority of this is in ROW, but with India and China still at about 1 billion tonnes per annum each, effectively from residual coal use.

- Cement use rises to about 5 billion tonnes per annum by mid-century, with abatement via CCS not happening until the second half of the century. One tonne of cement produces about half a tonne of processed carbon dioxide from the calcination of fossil limestone.

- Land-use carbon dioxide emissions have been assessed by many organisations, but I have used numbers from Oxford University's trillionth tonne assessment, which currently puts it at some 1.4 billion tonnes per annum of carbon (i.e. ~5 billion tonnes carbon dioxide). Given the NDC of Brazil and its optimism in managing deforestation, I have assumed that land-use emissions decline throughout the century, but still remains marginally net positive in 2100.

- I have not included gases such as methane that only reside in the atmosphere for one to two decades. These contribute more to the rate of temperature rise than to the eventual outcome, provided of course that by the time the end of the century is reached emissions of these gases have been successfully managed.

- Cumulative emissions stand at about 600 billion tonnes carbon in early 2016 according to the trillionth tonne calculator published by Oxford.

The result of such an assessment sees carbon dioxide emissions peaking at 47 billion tonnes per annum in 2030 and declining thereafter. By 2050 emissions are at 37 billion tonnes, then 8 billion tonnes in 2100. With an extended trajectory it is possible to assess the global cumulative emissions and temperature. The all-important cumulative emissions top out at just below 1.4 trillion tonnes carbon.

Using my assumptions, the trillionth tonne point, or the equivalent of 2°C, will be passed at around 2050, some 11 years later than the 2038 date indicated on trillionthtonne.org, the Oxford University Department of Physics website, at the time of COP21. My end point is therefore equivalent to about 2.8°C, well below 4 +°C, but not where it needs to be. The curve has to peak within the period of the first NDCs, i.e. before 2030, for global cumulative emissions to be substantially lower.

This is because time is the number one enemy of an accumulating system. In an accumulating system with a fixed total not to be breached, each additional year of emissions at current levels effectively cuts into the time available to bring emissions down, meaning that net-zero emissions must be reached even earlier.

The UNFCCC press release's reference to 2.7°C is plausible if subsequent NDCs follow through with continued reductions so that emissions start to fall after 2030, reaching net-zero late in the century or early next century. But this is the most optimistic assessment possible.

A similar assessment of the NDCs by the MITJP led to a different outcome. That analysis took the NDCs at face value and assumed that they would be implemented as indicated, but without significant follow-through reductions and certainly not achieving net-zero emissions around the end of the century. Arguably this is at the other end of the optimism spectrum. The MITJP assessment gave an outcome in the range of 3.1–5.2°C, which is significantly above the outcome indicated by the UNFCCC in its press release.

Even with a five-year review period built into the Paris Agreement, can the outcome in 2030 or 2035 differ significantly from this outlook? Will countries that have mapped their emissions through to 2030 change this part of the way through or even before they have started along their intended pathway? One indication that this might be the case comes from China, where a number of institutions believe that national emissions could peak well before 2030.[23]

RETHINKING KAYA

There is great optimism around emissions reduction, but much of this optimism is based on flow-problem thinking, rather than treating carbon dioxide emissions as a stock problem. Today there is a huge focus on renewable energy and energy efficiency as solutions for reducing carbon dioxide emissions. Clearly the strategy has been operating for over 20 years with emissions going only one way. Up.

A question that should be asked is, why have so many people fixed their hopes on renewables and energy efficiency?

This focus as a solution to rising carbon dioxide emissions possibly stems from the wide dissemination of the Kaya Identity, developed in 1993 by Japanese energy economist Yoichi Kaya. He noted that:

$$\text{Carbon dioxide emissions} = \text{People} \times \frac{\text{GDP}}{\text{Person}} \times \frac{\text{Energy}}{\text{GDP}}$$
$$\times \frac{\text{Carbon dioxide}}{\text{Energy}}$$

Or in other words:

$$\text{Carbon dioxide emissions} = \text{Population} \times \text{Development}$$
$$\times \text{Efficiency} \times \text{Energy Source}$$

Therefore, by extension and summing (the $\sum$ symbol) this up over many years (where k is the climate sensitivity and ΔT is the warming of the climate system):

$$\Delta T = k \sum_{1750}^{2100} \{\text{Population} \times \text{Development} \times \text{Efficiency}$$
$$\times \text{Energy Source}\}$$

In most analyses using the Kaya approach, the population and development terms are bypassed. Population management is not a useful way to open a climate discussion, nor is any proposal to limit individual wealth or development, i.e. Gross Domestic Product (GDP) per person. The discussion therefore rests on the back of the argument that because rising emissions are directly linked to the carbon intensity of energy (Carbon Dioxide/Energy) and the energy use per unit of GDP (Energy/GDP or efficiency) within the global economy, lowering these by improving energy efficiency and deploying renewable energy must be the solutions.

But the Kaya Identity may simply be describing the distribution of emissions throughout the economy on an annual basis, rather than the real economics of fossil fuel extraction and its consequent emissions. When an approach is taken that includes extraction economics, a different understanding of the problem emerges.

A mineral such as coal can be picked off the ground and exchanged for money based on its energy content. The coal miner will continue to do this until the accessible resource is depleted or the amount of money offered for the coal is less than it costs to pick it up and deliver it for payment. In the case of the latter, the miner could just wait until the price rises again and continue deliveries. Alternatively, the miner could aim to become more efficient, lowering the cost of pickup and delivery and therefore continuing to operate. The fossil fuel industry has been doing this very successfully since it first began. The fall in global coal prices from 2012 to the end of 2015 saw a number of coal mines close, but the partial recovery in 2016 saw some mines reopening — for example, the Glencore Collinsville mine in Australia.

The impact of fossil fuel extraction on the climate (ΔT) is therefore a function (f) of the total fossil fuel resource available in the sub-surface expressed in terms of carbon, and the price driver for extraction being the difference between the prevailing energy price and the cost of extraction. The price driver rises as extraction becomes more efficient. This gives the following:

$$\Delta T = kf\{\text{Fossil resource[Gt carbon]}, \\ (\text{Energy price} - \text{Extraction cost})\}$$

The equation simply establishes the amount delivered from the resource, not how efficiently it is used, when it is used, how many wind turbines are also in use or how many people use it. The development of global fossil resources and its subsequent use over the ensuing years is a closer measure of the reality of the climate problem. The larger the resource base that is developed globally, the higher the likely eventual concentration of carbon dioxide in the atmosphere. A version of this dynamic is playing out in the United States today.

As was to be expected, the United States submitted an NDC that indicated a 26–28% reduction in national emissions by 2025 relative to a baseline of 2005. This is an ambitious pledge, and highlights the changes underway in the US economy as it shifts towards more gas, backs out the domestic use of coal, improves efficiency and installs renewable generation capacity. So far, US national emissions reporting indicates that their 2020 target (a 17% reduction) is being progressively delivered.

Direct emissions represent just one strand of US emissions. Some might argue that national emissions reporting should also include embedded emissions within imported products, although this would introduce considerable calculation complexity into the estimate.

But a look at US carbon commitment to the atmosphere from a production standpoint reveals a different emissions picture. Rather than seeing a drop in US emissions since 2005, the upward trend that has persisted for decades is continuing. In the case of measured direct emissions, reduced coal use and improved efficiency are driving down emissions. But in terms of extraction, additional coal is now being exported and the modest drop in coal production is being more than countered by increasing oil and gas production. Total carbon extraction is rising, up 15% from 2005 to 2015.

While it is unlikely that national emission inventories will start being assessed on such a basis, it does throw a different light onto events. During a visit to Norway in early 2016 I was interested to hear about national plans to head towards net-zero emissions, but for the country to maintain its status as an oil and gas exporter. This would be something of a contradiction if Norway were not such a strong advocate for the development of CCS, a strategy that will hopefully encourage others to use this technology in the future.

The above may also mean that the energy price has to fall very low for extraction to slow down. Of course, that is where renewable energy can play an important role, but the trend to date has been for total energy system costs to rise as renewable energy is installed. A further complication arises in that once a mine or well is operating and all the equipment for extraction is in place, the energy price has to fall below the marginal operating cost to stop the operation. This can be extremely low for some operations where even the use of energy by the production facility may not be contributing to the operating costs (the extraction facility uses energy generated from the resource itself).

At such low-energy prices the extraction facility owner may go bankrupt in the process as capital debt is not being serviced, but that still doesn't necessarily stop the facility operating. Instead, it may be sold and the lost capital written off.

Looking at this on a macro scale, where thousands of extraction facilities are already in existence all over the world and are designed to last long enough to extract the resource they are sitting on, the task of limiting global emissions becomes a very difficult one.

Current global proven reserves of hydrocarbons[24] will release some 0.9 trillion tonnes of carbon when used, irrespective of how efficiently the hydrocarbons are used, how many wind turbines are built in the interim or how many green jobs are created while they are being used. In combination with cement production and continued land-use change, this will take the cumulative carbon towards 2 trillion tonnes by late this century, with the likelihood of a temperature increase of well over 2°C.

Opting to leave these reserves in the ground permanently (i.e. forever) so as not to contribute to the ocean/
atmosphere carbon stock will happen only if alternative
energy sources are developed that out-compete them for
every service they provide, without subsidy or support,
24 hours a day, 365 days a year.

THE NEED FOR CARBON CAPTURE AND STORAGE

Another way forward is to recognise that many economies around the world will choose to continue using the
resources that they have, and therefore the focus should
be on the development and deployment of CCS. This is
the capture of carbon dioxide from the combustion process and returning the carbon back to the sub-surface
(geosphere) instead of allowing it to accumulate in the
ocean/atmosphere system.

CCS uses existing processes and technologies available
in the oil and gas industry to capture and compress the
carbon dioxide from the combustion of fossil fuels. It is
sequestered deep below the earth's surface, one to three
kilometres, within geological formations suitable for permanent storage.

Carbon capture and storage is a technology that is specifically designed to counter the issue of accumulating
carbon dioxide in the atmosphere, rather than relying on
tangential approaches such as energy efficiency standards
and renewable energy directives. These latter policy
approaches may affect the short-term consumption of fossil fuels in one region for a limited time period, but they
offer no guarantee of permanent reductions nor do they

deliver a guarantee of a lower cumulative stock of carbon over time. In other words, the fossil fuel that they displace locally may get shifted geographically (used elsewhere) and/or temporally (used later) such that the same accumulation of carbon eventually results.

However, a tonne of fossil fuel consumed with emissions captured and stored is a very different proposition.

The IPCC in their 5th Assessment Report (Working Group III, Mitigation) demonstrates the economic benefit of CCS through scenario analysis. For scenarios ranging from 450 ppm CO_2eq[25] up to 650 ppm CO_2eq (i.e. a 2°C pathway to a 3°C pathway), consumption losses and mitigation costs are given through to 2100, with variations in the availability of technologies and the timing (i.e. delay) of mitigation actions.

Particularly for the lower-concentration scenario (430–480 ppm) the IPCC findings highlight the importance of carbon capture and storage. For the 'No CCS' mitigation pathway, i.e. a pathway in which CCS isn't available as a mitigation option, the costs are significantly higher than the base case that has a full range of technologies available. For higher-concentration scenarios higher costs are still seen, but not to the same extent. The IPCC analysis underpins the argument that the energy system will take decades to see significant change and that therefore, in the interim, CCS becomes a key technology for delivering an outcome that approaches the 2°C goal but without limiting energy system growth. For the higher-concentration outcomes, immediate mitigation action is not so pressing and therefore the energy system has more time to evolve to much lower emissions without CCS — but with the consequence of elevated global temperatures.

CCS has the potential to address carbon dioxide emissions on a scale equal to its production and at a cost that appears more than manageable by society. It is also a technology ready to deploy at scale as a CCS installation is a combination of existing oil and gas industry technologies pieced together in a different configuration. Most importantly, it fits the stock-model thinking, which means that this particular solution matches the nature of the problem itself.

But CCS is struggling politically to gain the necessary funding and momentum. Some policy makers see it as experimental, others worry about the fate of the stored carbon dioxide and yet others believe that it isn't necessary to solve the climate issue. So just to be clear, CCS isn't experimental — it's a reworking of existing oil and gas technologies. With CCS, the carbon dioxide will remain trapped deep below the earth's surface as oil and natural gas has remained trapped for millions of years and, yes, CCS will be necessary to solve the climate problem.

However, there are no large-scale CCS power generation plants operating in the world today (but there is a successful 100 MW coal-fired power station operating with carbon capture in Saskatchewan, Canada) and only a handful of industrial-emission CCS facilities, with a number of others under construction. New thinking and impetus is needed to ensure that CCS becomes central to climate policy development, rather than having to compete with the long list of other objectives that seem to prevail.

The importance of CCS is highlighted by another critical element of the climate issue, the need to eventually reach net-zero emissions. Until this goal is reached, carbon dioxide will continue to accumulate in the atmosphere, driving up the surface temperature. With some element of fossil fuel use almost certain to remain in the energy mix until well into the 22nd century, net-zero cannot be achieved without CCS.

CCS would also be required should a late 21st century or 22nd century strategy of carbon dioxide capture from air be implemented, designed to start reducing the atmospheric concentration.

And yet, as noted, CCS is struggling for recognition, even at UNFCCC conferences. As international delegates meet in plenary halls at the annual COP, the largest UNFCCC event to push the climate agenda, the side events and external conference programmes roll on in the surrounding halls and venues. It is through these debates that participants can meet and discuss subject related to climate change, such as fossil fuel use, national emission projections and carbon pricing.

Given that these are meetings about climate change, it might be expected that attendees would be interested in hearing about CCS, but it turns out to be a hard sell, even here. At COP 19 in Warsaw in November 2013, the problem began with the venue itself, where the meeting room banners were each headed with a particular environmental discipline. Having scoured the building, I found banners for Energy Efficiency, Renewable Energy Sources, Air Protection and Water & Wastewater Management. CCS didn't get a mention.

At the first post-Paris COP held in Marrakech, low-carbon technology featured high on the agenda, to the extent that a conference within the conference was held on this subject. CCS featured briefly on the final day, but even the role of information technology in reducing emissions was given greater prominence. Apparently IT services can help leverage greater efficiency across society, with services such as Uber and Airbnb. But IT has become amongst the largest energy-consuming sectors on the planet. And no account is made of the fact that a service such as Airbnb simply encourages even more people to use air travel.

I could take you to any number of climate-change presentations, speeches, dinner conversations and panel discussions but you would hear little mention of CCS. Where it does feature is when organisations such as the Global Carbon Capture and Storage Institute (GCCSI) hold events, but this might be expected as GCCSI is an institution designed specifically to promote the use of CCS. Otherwise the focus today is typically on energy efficiency and renewables, both important energy topics, but not necessarily reliable mitigation strategies. Further, the focus on renewable energy is largely an electricity focus. But electricity makes up just 20% of final energy, with 80% coming from direct use of oil products, natural gas and coal to deliver energy services.

When CCS does crop up, it is increasingly phrased as CCUS, with the U standing for 'Use' (carbon capture, use and storage). Back in Warsaw, the UNFCCC Executive Secretary referred to it as CCUS. In another forum, one participant talked about 'commoditising', carbon dioxide to find a range of new uses. The problem is that carbon dioxide can't be used for much of anything, with one modest but important exception. The largest use today is for enhanced oil recovery[26] (EOR). In the United States, a small but growing industry has now sprung up around CCUS because of this. Originally built on the back of natural carbon dioxide extracted from the sub-surface, the industry now pays enough for carbon dioxide that it can provide support to carbon capture at power plants and other facilities (usually with some capital funding from the likes of the Department of Energy). This has helped the United States establish a CCS demonstration programme of sorts with seven[27] operating projects.

There are other minor industrial gas uses — the fizz in soft drinks, for example. There is some scope for vegetable greenhouses — for example, a project in the Netherlands that provides refinery carbon dioxide to Rotterdam greenhouses for enhanced growing. There is also a technology that quickly absorbs carbon dioxide in certain minerals to make a new material for building. But all of these are very small in scale.

The problem is that carbon dioxide is the result of combustion and energy release. Any chemistry that turns it into something useful again requires lots of energy. Nature achieves this by using sunlight. Even if such a step were possible, it would hardly change the carbon dioxide balance in the atmosphere since no bio-process materially changes the overall balance in the atmosphere. Only sequestration, either natural or anthropogenic, changes that balance.

Accounting is also problematic for carbon dioxide use. For the use of carbon dioxide to be an effective mitigation measure, the material that is made from it must also be stored. This would be accomplished by increasing the total stock of the material in use at any one time.

Returning to the example above, say that homes are built with the new material. The starting point would be zero homes, but in a decade or so there might be 50 million homes constructed. Even if the homes are eventually torn down (and the carbon released), so long as the total number of such homes in use continues to increase, more and more carbon is stored. The issue here is that the total stock has to be maintained for a very long time (at least a century or more) for carbon dioxide use to approach CCS equivalence.

As populations grow and development proceeds, the stock of all goods in circulation has generally increased, even as old items are removed and new ones added. There are more buildings than ever before, more stuff in the buildings and more machines such as cars, ships and planes. All of these could be potential carbon stocks for century-long storage. But there is a corollary to be aware of, i.e. winding down the global stock of a certain item will result in the stored carbon being returned to the atmosphere.

Although a CCS facility would be comprised of a set of processes that are in widespread use in the oil and gas industry, including the final step of carbon dioxide injection into a suitable deep geological formation for storage, there is an acute global scarcity of fully functioning end-to-end CCS facilities in operation. This undermines policy makers' confidence in this solution. The lack of CCS facilities is hardly a surprise. Such facilities require considerable capital investment to construct, cost money to operate and deliver no tangible benefit to the operator, unless there is a sufficiently high carbon price operating in the energy system. For the first generation of facilities using current technologies designed for other purposes, this comes in at around $100 per tonne of carbon dioxide, although there are some limited opportunities for lower cost CCS, such as the capture of pure carbon dioxide released in the fermentation step of ethanol manufacture. As carbon dioxide storage and transport infrastructure is established, the cost for the next generation of facilities will also be lower.

But it wasn't always the case that CCS was the climate-change wallflower. At the Gleneagles Summit in 2005, G8 leaders committed to work to accelerate the deployment and commercialisation of CCS technology. This was followed by a further recommendation at the 2008 G8 Summit in Japan that 20 large-scale CCS demonstration projects should be launched by 2010. Similarly, the EU Heads of Government declared the need for 12 projects in the EU. But the implementation story is a very different picture.

The most successful projects to date can be found in Canada and the United States where fiscal incentives, grants and a robust carbon dioxide demand on the back of the extensive use of EOR in the oil industry have allowed a number of projects to get across the line. All have demonstrated that the technology is feasible, scalable and ready to go. The projects have also shown that future facilities can be delivered at lower capital and operating cost.

These early steps begin to lower the risk of the technology for future business investment, bringing some level of certainty to expected capital expenditure and ongoing operating costs in particular. For CCS there are other benefits as well.

- A demonstration programme that comprises several projects begins to establish some infrastructure, which in turn lowers the cost for additional projects. Perhaps the best current example of this is in Rotterdam, where there are proposals for a carbon dioxide pipeline loop in the industrial area of the city, connected to offshore storage in the North Sea. Given the large number of installations in the area, in combination with those further up the Rhine, real synergy is possible.

- In some regions (and Europe is one, Germany in particular) there are growing issues with public acceptance of carbon dioxide storage, even though there is little ongoing storage underway. There are fears that carbon dioxide may leak into the water table or into enclosed locations at the surface. A demonstration programme offers the opportunity to allay the fears related to storage and show that once the gas is sequestered some three kilometres below the surface in a suitable, properly monitored geological formation, it stays there.

- Demonstration improves best practice in the operation of a CCS facility and related storage, particularly in a major grid dispatch situation.

Demonstration efforts in the EU took off in 2008. With the support of companies looking to develop the first generation of CCS projects in Europe, the EU implemented a facility to leverage the prevailing carbon price and offer a significant capital grant to about a dozen proposed projects. Given Shell's interest in seeing CCS develop, I was one of the initiators of this idea. The concept of using allowances from the EU Emissions Trading

System (EU ETS, a cap-and-trade system) as a CCS support mechanism had been around for some time, but the idea of using them as part of a clearly defined project mechanism started one Sunday afternoon at my kitchen table when I wrote out a concept paper describing this, including some rudimentary legislation text for our Brussels office to share with some members of the European Parliament — but mainly with one particular stalwart for CCS and rapporteur of the CCS Directive, the UK Liberal Democrat MEP Chris Davies. The idea gained momentum, was reshaped as it progressed and eventually ended up as a real mechanism in Europe.

Known as the NER300, it allowed for auctioning of 300 million allowances from the New Entrant Reserve (NER) to generate the grants (totalling €9 billion at a price of €30 per tonne). Given that the prevailing carbon price was also high enough to underpin the ongoing operation of a CCS plant, some twenty or so projects were submitted from across Europe seeking access to the grant money.

Unfortunately, the commencement of the NER300 also marked the last the EU would see of CCS. The price of EU allowances started falling in late 2008 and by 2016 had not recovered. Although the NER300 went on to support several renewable energy projects, the size of the grants that emerged from a €6 market weren't relevant for CCS.

As it turns out, getting the carbon price to a level that triggers real technological change, such as the implementation of CCS, has been much more challenging than expected.

4

WHY CARBON PRICING MATTERS

THE POLICY LANDSCAPE

Over recent years, the focus of energy policy for emissions reductions has been a combination of targets, energy mix mandates, efficiency drives and various attempts at carbon pricing. Energy and infrastructure policies that are also claimed to address climate change include renewable energy targets, efficiency goals, public transport initiatives, cycling schemes and urban redevelopment.

But are these the right types of policies for solving the carbon dioxide emissions problem? There is no doubt that such approaches have gained traction and wide support from policy makers, but in many instances they are the result of a desire to solve a broad range of topical issues, ranging from energy security and energy access to jobs and economic growth. There is an underlying assumption that because each of these is related to domestic emissions reductions or near-zero emissions, they must offer a solution to the real elephant in the room, the rising levels of carbon dioxide in the atmosphere. But this assumption may not be the case.

These policies assume that addressing climate change depends on managing the rate of emissions from the global economy, sometimes on an absolute basis but often on a relative basis, e.g. relative to GDP. All these policies assume that the carbon dioxide issue is a flow problem. But we have already established that this isn't the case.

Some people seem to take the view that energy efficiency is a viable emissions reduction strategy in itself, and therefore interchangeable with technologies such as CCS, which removes carbon dioxide from the accumulating stock.

An example of this assumption comes from looking at the closure of older, less-efficient coal-fired power stations in China since 2010, largely in response to deteriorating air quality in cities such as Beijing. Many of the early power stations were built too close to the cities that they serve. Chinese coal use declined in 2014 and 2015 as a result of these closures, but can this shift be translated into a long-term reduction in atmospheric carbon dioxide? The likely reality is that the same coal is being used more efficiently in newer and cleaner power stations to generate even more electricity, but with less impact on air quality. Most of these units are in other parts of China, but the drop in Chinese demand could also spur new coal-fired power-station construction in other parts of Asia. Even if it could be shown that China had actually reduced its longer-term coal consumption as a result of this change, is there certainty that this also means permanent closure of the coalmines feeding these power stations? If the mines are eventually depleted over a longer period, the same amount of carbon is still released leading to the same amount of warming.

This is not dissimilar to government attempts to cut deficits. Many countries have seen deficits rise in absolute terms since the idea of deficit spending was first introduced. Yet successive governments have all implemented efficiency drives to reduce the deficit and have claimed some success. The problem is that the reductions are often set against projected spending rather than current spending, so a reduction can be claimed at the same time as the reality of an absolute increase in spending. As such, the total deficit continues to rise. Real deficit reduction will probably only come with major structural changes in government policy (welfare, defence, etc.), but these are much more difficult to implement. At least with government spending there is a relief valve of sorts: the economy can grow and therefore the cumulative deficit can shrink as a fraction of GDP. Unfortunately, there isn't such a relief valve in relation to the atmosphere.

Even so, organisations such as the IEA also adopt this line of thinking in their annual outlook reports. These reports project what business-as-usual emissions would be by some future date and then argue that a focus on energy efficiency could reduce this, effectively claiming an emissions reduction. Nevertheless, emissions continue to rise. This reasoning appears to show energy efficiency as the most important contributing factor to change. Yet, in reality, the original projection represents a situation that may never have occurred. Business-as-usual requires improvements in energy efficiency to drive growth. If energy efficiency really is a route to lower emissions, then it needs to pass one clear test: which known fossil fuel resource will be left in the ground, or which proposed extraction project will be shelved because of this? Only then are cumulative emissions potentially impacted.

One unintended consequence of energy efficiency policy can be to exacerbate the emissions problem. In the worst case scenario, an energy efficiency improvement in the power generation supply chain can incentivise the resource holder (e.g. coal mine) to expand the resource base and therefore increase the potential tonnes of carbon that will be released into the atmosphere.

This is because the electricity provider can now afford to pay more for the coal, which may in turn expand access to new coalfields. This drives up long-term warming, even if the rate at which carbon dioxide is emitted falls in the short term. A similar outcome can result from efficiency improvements in fossil fuel production itself. As extraction costs fall, access to the resource base expands for the same energy market price.

Energy efficiency is a key driver for development, primarily through the reduction in cost of energy services. This increases access and availability of those services and therefore spurs development. Arguably, it has been the single most important element of the industrial revolution, underpinned by key inventions along the way. But we now seem to have decided that this is critical to solving climate change. Is it?

Exciting new technologies such as LED lighting are purported to reduce energy use, therefore emissions. Today, thanks to LEDs, it's not just the inside of buildings that are lit up at night, but the outsides as well. In cities such as London and New York, entire buildings glow blue and red, lit with millions of LEDs that each use a fraction of the energy of their incandescent counterparts — or would do if incandescent lights had been used to illuminate cityscapes on the vast scale seen today. Similarly, the falling cost and increasing efficiency of LEDs has driven them into the world of advertising, with billboards and signs moving from paper to electronic. The response to lighting efficiency has been to consume more lighting services and potentially more energy, thus possibly increasing emissions. Yet again human nature has gotten us into trouble.

An analysis from Sandia National Laboratories, a US energy research centre, looked at this phenomenon and concluded that increases in lighting efficiency have been accompanied by an increase in demand for energy used for lighting that nearly exactly offsets the efficiency gains — essentially a 100% rebound in energy use.

Similar effects have been observed in the transport sector, where improvements in the cost of mobility services expand their use — just look at how accessible and affordable air travel has become over the last few decades. Information technology (IT) is often claimed as an enabler of emissions reduction through efficiency, but in the air transport sector, IT-enabled efficiency has allowed low-cost carriers to fill every seat, dramatically lower the cost of travel per passenger and spectacularly expand their businesses. Similarly, in the home, as efficiency measures improve home heating, there is evidence that the average indoor temperature rises, while energy use doesn't fall as much as anticipated.

One route to solving the climate issue is to stop using fossil fuels and leave them in the ground forever. For this to happen, something else has to replace them at lower cost, not just sometimes, under the right conditions, but all the time and without subsidy.

Renewable energy commands considerable subsidies, either directly in the form of mechanisms such as feed-in tariffs, but also indirectly in the form of reliable backup (e.g. natural gas) for when the renewable source is not available. The back-up source is rarely, if ever, included in the cost of providing renewable energy.

Even as the real cost of providing uninterrupted renewable energy falls, there remains the issue of energy infrastructure turnover. Energy assets tend to have longevity, with coal-fired power stations lasting for decades. Some of the facilities still generating electricity in the United States in 2016 were built as early as the 1950s. Even the eventual closure of a facility, perhaps because of its replacement with a wind turbine array, doesn't necessarily mean that the associated coal mine will also be shut down. Again, there is evidence of this in the United States where the closure of many coal-fired power stations in the decade from 2006 has resulted in displacement of at least some of that coal to the export market, rather than permanent closure. This coal continues to add to the cumulative emissions of carbon dioxide as it is consumed.

In the global market, demand for energy continues unabated. The addition of a bit more US coal into the global market may lower the price slightly, but only to the benefit of the next person in that long line of people without sufficient access. Coal remains an attractive option because it requires minimal infrastructure and technology investment to utilise.

DERIVATIVE APPROACHES

Renewable energy advocates are, not surprisingly, incredibly optimistic about future deployment rates of solar and wind technologies. They have good reason to be. But one of the comments I come up against from time to time is, "The oil and gas industry has only got 20 years." This doesn't just come from enthusiastic climate campaigners, but sometimes from thoughtful professionals in disciplines related to the climate issue.

It's hard for anyone who has worked in this industry, as I have, to imagine scenarios in which fossil fuel use vanishes from the planet during the next two decades. This isn't entirely down to vested interest, but because of the vast scale, complexity and financial base of the industry itself. The fossil fuel industry is the legacy of over two centuries of work, at a cost of trillions (in today's dollars); at the time of the Paris Agreement it provided over 80% of primary energy globally (with that demand nearly doubling since 1980), with market share hardly budging. Global energy demand may well double again by the latter part of the century.

So why do people think that all this can be replaced in a relatively short space of time? The enthusiasm stems from the current rapid growth that is clearly visible for new renewable energy technologies such as solar PV, in some instances as high as 50% per annum.

But the more realistic path is one in which a burst of growth is followed by much more regulated expansion limited by resources, finance, intervention, competition and a variety of other factors. This is how energy systems tend to behave — they don't continue to grow exponentially. Historically, there are many examples of rapid early expansion to the point of materiality (typically ~1% of global primary energy), followed by a long period of growth to a level representative of the energy source's economic potential. Even the first rapid phase takes a generation, with the longer growth phase stretching out over decades.

Addressing the climate issue with a strategy that hinges on increasing the supply of renewable energy and improving the efficiency of energy use is unlikely to yield results, particularly when the problem, as we have seen, is primarily defined by the existence of the fossil fuel resource base itself.

Rather than tackle the issue head on, the assumption is made that by doing something else that is related, the original problem might solve itself or go away. It won't. This is what could be termed a derivative approach. The climate issue is about the release to atmosphere of fossil carbon and bio-fixed carbon on a cumulative basis over time, with the total amount released being the determining factor in terms of peak warming (i.e. the 2°C goal). It isn't about how many wind turbines or solar installations can be built over the coming decades. One possible outcome is that the world will have lots of wind turbines, without seeing a reduction in carbon dioxide emissions.

Energy efficiency policy is also a derivative approach in that it attempts to limit the rate at which energy use increases, which is yet another step further away from directly reducing emissions. Energy efficiency policy is influenced by other underlying factors, such as how often each of us buys a new car or replaces our household appliances and how often manufacturers introduce new, better performing models. Efficiency policies aren't even designed to deal with the immediate rate of energy use; this is dictated by the existing infrastructure, for example, the number of cars on the road and appliances in the home. Efficiency tackles the change over time as old infrastructure — your car, for instance — is disposed of and replaced.

Focussing on renewable energy deployment and efficiency is an important and cost-effective energy strategy for many countries, but as a global strategy for tackling cumulative carbon emissions, it falls far short of what is necessary.

But this doesn't have to be the end of the story. Returning to the earlier equation linking fossil fuel resource availability with eventual warming of the climate system, the balance needs to be tilted to make extraction of the resource appear less economically attractive. The obvious ways are (1) to lower the energy prices or (2) to raise the extraction cost. But with global energy demand continually rising and with efficiency helping to drive down extraction costs, neither of these seems like a winning proposition. Instead, another term has to be introduced into the equation — the carbon price.

The climate change issue argues that global anthropogenic emissions of carbon dioxide return to net-zero levels, ideally within this century if the goal of limiting warming of the climate system to well below 2°C is to be achieved. Implementing public policy to deliver a cost for emitting carbon dioxide as part of the energy economy is arguably the single most important step that can be taken to achieve this objective. It is where governments started the climate journey nearly 20 years ago, but not where they find themselves now.

THE ROLE OF A CARBON PRICE

A carbon price is a cost put on the emission of carbon dioxide into the atmosphere from the use of the fossil fuel after it is extracted. It must be implemented by the government through legislation. Some governments are already doing this — for example, in the Canadian province of British Columbia there is a tax of C\$30 for every tonne of carbon dioxide emitted from fossil fuel use. For the motorist, this adds about 7 cents to the cost of a litre of fuel.

With a carbon price now in place, the climate equation changes as follows:

$$\Delta T = \ kf\{\text{Fossil resource[Gt carbon]},$$
$$(\text{Energy price} - \text{Carbon cost} - \text{Extraction cost})\}$$

The idea that a carbon price is needed can be traced back to the work of Arthur Cecil Pigou, a University of Cambridge economist who published *The Economics of Welfare* in 1920. In this book, Pigou introduced the concept of externality and the idea that external problems could be corrected by the imposition of a charge. By 'externality', Pigou meant an indirect economic impact of an activity that happened outside the immediate system where the activity was underway. The externality concept remains central to modern welfare economics and is at the heart of environmental economics.

In the case of climate change, the externality is the release of carbon dioxide into the atmosphere and the future social and economic impact caused by the consequent increase in the surface temperature of the planet. Externalities can be both positive and negative, but in the case of carbon dioxide emissions, it is considered negative. Pigou argued that the activities associated with a negative externality should be penalised to the extent of the impact, such that their real economic value can be assessed. This penalty is widely known as a Pigouvian Tax. Alcohol taxes are Pigouvian; so are taxes on cigarettes.

When the carbon price is high enough to offset the profit from the resource extraction, then the process will stop, but not before. This introduces a further dilemma, the overriding need to supply energy to meet global demand.

If a carbon price is in place but renewable sources of energy don't grow fast enough to meet global demand or can't replace the myriad uses for fossil fuels, then energy prices have to rise to shift the extraction economics back again. This may impact on growth and development prospects. Yet another piece of the puzzle also needs to fall into place — CCS.

Today, confronted with a variety of energy policy objectives, including energy security, the cost of energy and the need to reduce carbon dioxide emissions, many governments have opted for a common suite of solutions in response — that is, solutions that appear to address all the issues simultaneously. This might include very legitimate approaches such as renewable energy targets, efficiency goals, nuclear power investment and subsidies for electric vehicles. But in the case of the climate issue and given these solutions' divergence from the specific carbon emissions problem to be solved, they may not present an effective response. Even when local success is delivered, these and other measures can still prove ineffective against the primary global objective of limiting cumulative carbon emissions to some 1 trillion tonnes.

Similarly, actions directed at attempting to limit the use of fossil fuels in a particular industrial facility, or even in vehicles, may not deliver a comprehensive outcome at a global level. Another strategy is to try to drive down the price of energy with new supply options such as wind and solar PV in order to wean us off fossil fuels. But as already shown, success at scale is not guaranteed.

Alternatively, society could employ CCS as a means of continuing production while minimising exposure to the cost of emitting carbon dioxide. A few large-scale projects are now operating, such as the SaskPower Boundary Dam coal-fired power station that opened in October 2014 and the Shell Quest project that opened in Alberta a year later.

A cost associated with carbon dioxide emissions has been critical in establishing the first few CCS projects around the world. The original and best-known CCS project, Sleipner in Norway, is underpinned by a carbon tax imposed on the offshore oil and gas industry. In Saskatchewan, Canada the value that carbon dioxide has to the EOR industry has helped deliver the world's first coal-fired power station with carbon capture.

While the implementation of a carbon-emissions cost will initially trigger a range of activities throughout the economy, the only two activities that make a significant difference in meeting the challenge of carbon dioxide emissions are (1) reducing the extraction of fossil carbon, or (2) implementing carbon capture and storage. In this context, for the same provision of final energy, focussing on fuels such as natural gas rather than coal can reduce fossil carbon extraction.

GOVERNMENTS AND CARBON PRICING

Assessing what level would be needed for a carbon price is extraordinarily difficult because the full extent of future change is not known. The earliest well-documented attempt was by Professor Bill Nordhaus of Yale University in the 1990s. A number of commentators and institutions have attempted to do this since and results vary widely. According to data published on the US Environmental Protection Agency (EPA) website, they used a value of US$36 per tonne of carbon dioxide in 2015, although the EPA offers a complete table of values with a wider range depending on various factors. The US government has used these values, known as the 'social cost of carbon', to assess the benefits of certain regulatory actions to reduce carbon dioxide emissions. A social cost of $36 per tonne of carbon dioxide means that emitting an extra tonne of carbon dioxide today has the same impact on society's expected welfare as immediately reducing a representative consumer's consumption by $36.

In 2006 the UK Treasury asked their Chief Economist Nicholas Stern, formerly Chief Economist at the World Bank, to lead a major study on the economics of climate change. The 700-page *Stern Review on the Economics of Climate Change* has become the most widely known and most discussed report of its kind. Within it the social cost of carbon was estimated at US$85 per tonne, assuming continued business-as-usual emissions of carbon dioxide, or a much lower $30 per tonne for a pathway that sees atmospheric stabilisation of carbon dioxide at 550 ppm. The *Stern Review* proved controversial at the time as it proposed a very low discount rate when assessing future impacts, thereby raising the immediate social cost of carbon.

But rather than basing all our actions on a near impossible assessment of the social cost of carbon, the real cost of emitting carbon dioxide that has evolved in recent years is more typically a result of the policy process that determines what an emitter is required to do. This will include the type of policy framework put in place by government to attempt to reduce emissions and some medium-term objective associated with it. Not surprisingly a great deal of political horse-trading is involved in establishing the rules under which the framework will operate, but in my view the medium-term objective should at least be aligned with the long-term goal of limiting warming to well below 2°C.

One common objective should be to bring some economic order to what might otherwise be a chaotic and expensive process of reducing emissions within the economy. The tough question is always where to start and what to do first? Irrespective of the mechanisms employed, a successful policy framework should support the implementation of emission reduction projects and actions throughout the economy, with the lowest-cost result being delivered by encouraging the most economically attractive projects to be developed first and progressively moving towards the costlier projects. Carbon price based approaches are designed to do this.

Four such policy processes have evolved to deliver a carbon price:

Cap-and-trade (also referred to as 'emissions trading' in many regions): A notional cap is placed over some or all of the economy, meaning that total emissions under the cap cannot exceed a certain amount in each year. That amount usually declines over time, thereby driving emissions down. In some countries, such as India, where rapid development is underway, the cap may increase for some years but at a rate less than that associated with business-as-usual emissions growth. This limit is expressed in terms of allowances that the government issues, normally at some cost to the recipients, although under certain conditions they may be granted free of charge. The only obligation on an emitter operating within the system is to surrender one allowance for each tonne of carbon dioxide emitted, a process known as 'compliance'. The allowances are transferable through trade and have a value determined by the market as emitters seek to acquire allowances for compliance. The system forces a specific environmental outcome through the overall cap; in theory at the lowest overall cost to the economy because participants progressively implement projects in order of merit as the price adjusts to reflect the most economically attractive route forward. This policy mechanism has been implemented in several locations including Europe, California, Québec and South Korea and is also under development in China by the National Development and Reform Commission (NDRC).

Carbon tax: The government imposes a fixed tax on carbon dioxide emissions from specific sources in various parts of the economy. This may be where the emissions occur, or upstream of the actual emissions

(e.g. at the point of sale from a coal mine). Like a
cap-and-trade system, the carbon tax approach
requires measurement, reporting and verification of
carbon dioxide emissions across the sectors covered by
the policy. It is a relatively simple approach to
understand and implement, but requires significant
analysis with regards to the setting of the tax level in
order to achieve a specific environmental outcome.
This can only come from a clear understanding of the
opportunities present in the economy to reduce
emissions. Over time, the government may raise the
tax in order to drive emissions down. Examples of
carbon taxation currently operate in British Columbia
and Norway. The governments of South Africa,
Chile and Mexico are also implementing a modest
carbon tax.

Baseline-and-credit: The government establishes a
baseline emission for each sector, typically on a
unit-of-production basis, i.e. tonnes of carbon dioxide
per tonne of product made. This is also referred to as
an intensity-based approach in that it acts on the
carbon intensity of the activities it covers rather than
on the total emissions. The participants earn credits by
operating their facilities at lower emissions intensity
than the baseline, or surrender credits if they exceed it.
The credits are tradable and can be saved from one
year to the next (banked), as in the cap-and-trade
approach. Baseline-and-credit requires an accurate
assessment of the respective baselines across different
sectors, which can in turn be open to challenge and
debate within and between the sectors involved.

Because of the trade of credits, baselines should also represent an equivalent effort when comparing sectors, i.e. y tonnes of carbon dioxide per tonne of cement equivalent to x tonnes of carbon dioxide per tonne of steel. If these are not equivalent, economic distortion between sectors can result — in effect, state aid to a particular industry sector. Importantly, the environmental outcome in terms of absolute emissions is uncertain, as it depends on the level of production. However, the baseline will normally be reduced over time. Perhaps the best example of baseline-and-credit is the Alberta Specified Gas Emitters Regulation, which began operation in 2007.

Project mechanisms: A project is developed and emissions are compared with a baseline, which may represent best available technology or typical practice for a particular country. For example, if coal were the usual fuel for power generation, then this would be used to calculate the baseline. If the project emissions are lower than the baseline, credits are issued. These credits are tradable and may be bought directly by governments or used as compliance instruments in cap-and-trade systems in other jurisdictions. Like the baseline-and-credit approach, a project mechanism requires a high level of oversight, including baseline determination and measurement, reporting and verification. The mechanism typically requires an assessment panel of some sort, which may introduce a level of subjective decision-making into the process. As this is an opportunity-based mechanism rather than an imposed mechanism, the price is not relayed effectively through the economy. The Clean Development

Mechanism (CDM) under the Kyoto Protocol of the UNFCCC is the most widely used project mechanism in operation.

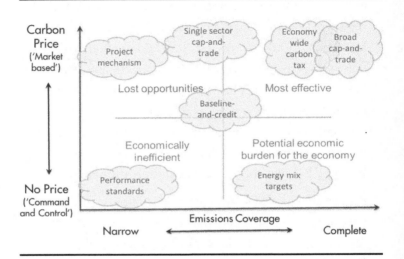

Comparing approaches and policies is difficult, but in general the various mechanisms can be rated as shown in the chart. The most effective approach is a widely applied carbon price across as much of the (global) economy as possible, delivered through a carbon tax or via a cap-and-trade system. Lost opportunities and inefficiencies creep in as the scope of approach is limited, such as in a project mechanism or with a baseline-and-credit approach; neither of which tackles fossil fuel use in its entirety. This inefficiency comes back to the relationship between fossil carbon resource use and eventual warming. If the price doesn't cover the full use of the resource and also span its potential use, extraction could still proceed until the resource is depleted, resulting in no change in the eventual level of warming.

The chart includes non-price-based mechanisms such as performance standards and energy mix targets, which tend to be less effective than a direct carbon price. Some policy makers favour these approaches because they can be designed to benefit certain sectors and also address other pressing issues (e.g. energy security), but their over-all effectiveness in reducing the carbon emission stock is uncertain and they do not necessarily deliver the lowest-cost opportunities first. They may also have unexpected consequences: for example, an efficiency measure may result in improved production and therefore lower costs but could also drive up overall emissions as demand for a cheaper product increases.

The most effective mechanisms for managing carbon dioxide emissions are those that directly impact the price of goods and services within the economy, which can also act as a deterrent to implementation. These price changes permeate the entire economy, creating a change in the market that begins to differentiate between various goods and services on the basis of their carbon footprint (or the total impact on emissions as a result of the purchase of the good or use of the service).

The carbon cost is initially experienced by the emitter or fuel provider (e.g. by paying a tax or purchasing allowances from the government) and may be passed through to the consumers of the product. Pass-through results in an increase in the absolute market price of most goods and services based on the carbon emissions within their respective supply chains, which in turn leads to the emergence of a new value ranking within the economy. This will influence the purchasing decisions of consumers. Products with a high carbon footprint will be less competitive, either forcing their removal from the market, or driving manufacturers to invest in projects to lower the footprint. The overall increase in cost for the consumer can be addressed by the government through recycle of the collected carbon revenue. This might result in a reduction in other taxes or charges that a consumer would normally bear.

The chart clearly shows carbon taxation and cap-and-trade competing for the top spot as the most effective mechanism for delivering the carbon price into the economy and driving lasting emission reductions. Both approaches work, so differentiating them almost comes down to personal preference, which can even be seen in the extensive academic literature on the subject where different camps lean one way or the other. My preference, perhaps influenced by my oil trading background, is to back the cap-and-trade approach. My reasons are as follows:

- The cap-and-trade approach delivers a specific environmental outcome through the application of the cap across the economy.

- Both instruments are subject to uncertainty: however, the cap-and-trade approach may be less subject to political change, while governments regularly change taxation policy as part of the annual budgetary process.

- The carbon price delivered through cap-and-trade can adjust quickly to national circumstances. In the EU it fell in response to the recession and perversely has stayed down in response to other policies (renewable energy goals) currently doing the heavy lifting on mitigation. In some sense, this is perverse because the other policies shouldn't be doing this job when a cap-and-trade is in place to do it more efficiently.

- Acceptance is hard to win for any new cost to business, but particularly when competitors may not be subject to that cost. The cap-and-trade system has a very simple mechanism in the form of free allowance allocation for addressing this problem, often faced by energy-intensive (and therefore carbon-intensive) trade-exposed industries. Importantly, this mechanism doesn't change the environmental benefit of the system or reduce the incentive to manage emissions because the allowances held by a facility still have opportunity value associated with them.

- Most carbon policies are being formulated at country or regional levels, rather than being driven by global approaches. Cap-and-trade systems are well-suited to international linking, leading to a more harmonised global price, while tax coordination is complex and politically difficult. Linking leads to a level playing field for industry around the world that fosters acceptance.

The economic effectiveness of both a carbon tax and a cap-and-trade system for carbon pricing means that countries and regions of all shapes and sizes have an implementation choice. For large, multi-faceted economies, the cap-and-trade system is ideally suited for teasing out the necessary changes across the economy and offering the many emitters considerable flexibility in implementation. Equally, for some economies or sectors where options for change are limited, the offset provisions that often feature in the design of various emissions trading approaches can offer a useful lifeline for compliance. Still, in certain economies, a direct tax may be the most appropriate approach. Perhaps this is for governance reasons related to trading, or a lack of sufficient market participants to create a liquid market or simply to encourage the uptake of a fuel such as natural gas rather than coal.

Many companies and governments manage the economic risk associated with the arrival of carbon-pricing policy through the application of a shadow valuation of carbon, the job of which is to mirror some external development. This is often loosely referred to as a carbon price, but more correctly is an *internal carbon value*. The internal value, when used by government, can be based on the social cost of carbon as part of a given stabilisation goal. In business, the internal value reflects the expected impact that medium to long-term government policy will have on a particular investment.

Some observers have concluded that an internal business approach should operate as an actual cost of carbon emissions within the respective business, such that the business behaves as if it were subjected to an external carbon tax operating at the same value. This would be done in the absence of an external carbon price driver, therefore acting as a stand-in for the lack of government action. But this is not what is happening or what is meant by an internal carbon value. There is an element of wishful thinking operating here, with some believing that internal carbon valuation can lead to widespread emission reductions as a major business-led initiative.

Rather, the internal carbon value also referred to as a shadow carbon value or carbon screening value is normally a mechanism used to manage the future regulatory risk that parts of the company or a future project may be exposed to. For example, if a certain investment is to be made, that investment is tested against a variety of possible future conditions, which could include an eventual cost incurred by the expected emissions of carbon dioxide. Although the project may not immediately be exposed to such a cost, the introduction of climate legislation may bring about exposure, which in turn could threaten the future viability of the asset. The application of a shadow carbon value applied when the investment proposal is being assessed allows the investor to reconsider the project, change the scope, modify the design or simply accept the level of risk and proceed.

This type of carbon value is often used interchangeably with the price that is derived from an external policy mechanism, but they are very different. The two that have a direct impact on emissions are the primary policy ones, i.e. the real and immediate cost imposed on carbon dioxide emissions that stems either from taxation or from an emissions trading system and results in the direct transfer of money within the economy.

A carbon emissions cost imposed through national legislation is designed to create winners and losers, with the winners being the ones that can best adapt their supply chains to manage the cost imposed on emissions and provide consumers with the most cost-effective and carbon-effective products and services. Major adjustments within the economy must be triggered to drive down emissions to net zero, albeit over many decades. This isn't about trimming the edges; it's about wholesale change.

Nevertheless, debate continues as to the most appropriate mechanism or approach for delivering such change. For example, given that the expected end state of the energy system is likely to see considerable deployment of CCS simply because a net-zero emissions society may not be possible without it, academics have argued that society should progress directly to this step and avoid the pitfalls and complexities of carbon pricing. This also avoids the wait while the carbon price rises to a level at which CCS is triggered on a very large scale.

The annual Forum held by the MITJP is an important event on the climate science and policy calendar, with excellent presentations and lively debate ensuing. The Forum held in Boston in October 2014 was no exception thanks to a discussion on two very different approaches to triggering the necessary mitigation of carbon dioxide emissions. But both were still market-based approaches with a pricing mechanism embedded within them.

The debate started with a presentation on cumulative emissions and the clear link to peak warming of the climate system, coupled with the critical role of carbon capture and storage. The challenge to the assembled group of climate scientists, economists and industry representatives was profound. Is, then, the simplest solution to the climate issue to enforce CCS, starting with a small amount for each tonne of carbon dioxide emitted, say 1−2%, but progressively increasing this throughout the century until 100% is reached? As the objective of CCS is to capture and store carbon dioxide, a mechanism that specifically rewards this activity rather than placing a cost on all the emissions to the atmosphere may be a more successful approach.

The development of a tradable CCS credit (where one credit represents one tonne of carbon dioxide stored) could be such a mechanism and be used to distribute the benefits of individual large projects amongst many, particularly in the early years when the capture and storage requirement from an individual emitter would still be small. It was argued in the Forum that this is economically more attractive than the widespread use of a carbon price mechanism, which would have to get to higher levels than current systems are offering to trigger meaningful CCS deployment.

In the case of CCS credits, which individually would reflect the cost of capture and storage, the actual cost for an emitter with a carbon storage compliance requirement would nevertheless be initially small. If this were to start in 2020 in a given jurisdiction at 1% and reached 15% capture and storage by 2030 (i.e. 100% by mid-2080s), the average cost over the period 2020−2030 to an emitter would be $8 per tonne of carbon dioxide, for the case where CCS credits trade at $100 each (on the basis that for the 1% case, the $100 cost for one credit spread across 100 tonnes of emissions is $1 per tonne of carbon dioxide emitted). This is about the level of the EU ETS in early 2015, which is unlikely to result in any CCS deployment at all.

It was further argued that an even more compelling reason for implementing the credit approach is that it would force CCS implementation right from the start, even on a local basis, which serves to address the accumulation issue immediately, rather than delaying implementation with measures that may ultimately make no difference to the eventual level of carbon dioxide in the atmosphere (e.g. local energy efficiency improvements).

The carbon-pricing economists at the MIT Forum responded to this, arguing that the direct pricing approach was more efficient; it would allow a range of other mitigation options to play out in the interim before CCS was needed, such as fuel switching from coal to natural gas. Debates such as this will doubtless continue, as will debates regarding the design and implementation of specific carbon pricing systems.

Broad implementation (after some period of pre-commercial demonstration) of CCS technology is going to require a higher carbon price than existing systems have ever delivered, but given the resistance to such a level appearing in the energy system it might be a very long wait indeed. Nevertheless, the flexibility and economic efficiency of a price-based mechanism are very desirable traits.

CARBON PRICING IN PRACTICE

Since 2005, as a result of a cap-and-trade system known as the EU ETS, a carbon price has been in operation across the European Union, with the addition of Norway, Iceland and Liechtenstein in 2008.

The real aim of a cap-and-trade system is to encourage investment in energy infrastructure so that emissions begin to fall within the economy and long-term lock-in, the result of continued construction of facilities such as coal-fired power plants, is avoided. This requires a price

that indicates such a requirement — perhaps around
$30 + per tonne of carbon dioxide for fuel-switching
away from coal and towards natural gas (but it depends
entirely on the relative prices of these commodities at any
point in time) and something over $100 + per tonne for
CCS (but lower in the future as this technology matures).
The policy framework that delivers the carbon price must
persist and project developers will need to have confidence that it is both there to stay and reflective of the real
abatement opportunities in the economy. Otherwise, they
won't invest.

For some high-emitting facilities, a persistent price in
the range $30–$100 is hard to live with. A tonne of coal
has traded in recent years for as little as $50 per tonne, so
a carbon dioxide price of, say, $75 per tonne will drive
the effective cost of that coal to around $200 per tonne,[28]
which is a higher price than coal has ever traded for. Of
course, a carbon price is specifically designed to impact
coal.

But what is the impact on an important building
material such as cement? The manufacture of cement
unfortunately results in a considerable quantity of carbon
dioxide emissions. Although the industry has made great
strides in improving the overall cement-making process,
its fundamental chemistry means that carbon dioxide is
emitted when limestone is heated, with the heating
process itself also being energy-intensive. A tonne of
cement might sell for US$100 but in its manufacture
(as clinker) can release about half a tonne of process carbon dioxide, so the price of cement rises significantly
under a carbon emissions cost regime that delivers a
meaningful price, irrespective of the source of energy used

in the calcination process. That may be tolerable in the case of a universally applied system, but the cement maker may still feel hard pressed to compete with alternative building materials.

In the early years of operation of the EU ETS, almost all allowances were granted for free to all emitters, largely to quell the issues outlined above and not disturb the status quo too much. This brought with it another set of issues — windfall profits. Despite being free at the point of allocation, the allowances still had value in the market, which meant that the emitter was exposed to the cost of allowances at the margin, either needing to buy some if there was a shortfall or having some to sell if emissions were lower than expected and there was an allowance surplus. Particularly in the case of electricity, where the marginal kilowatt hour generated effectively sets the market price for all the electricity, this meant that the cost of emissions quickly appeared in household bills even though the generator was receiving allowances for free. The end result was a windfall for the electricity producer. This issue is history because all of the allowance distribution for the electricity sector is now managed through government auctions; but in the early years of the EU ETS, it caused considerable friction.

Although auctioning has prevailed in the electricity sector, the same cannot be said for the industrial sector, where allowance allocation is still largely by free distribution. But this distribution isn't giving rise to windfall profits; rather, it is helping address a very different issue. A carbon price is normally passed through the supply chain where it impacts the cost of goods and services provided to the consumer. But if a competitor outside the cap-and-trade system is setting the market price for those goods and services, then pass-through isn't possible. The supplier inside the cap-and-trade system becomes competitively disadvantaged, and the supplier outside the system gains market share and may therefore see a rise in emissions from its manufacturing facilities. This is known as carbon leakage and is managed by distributing some or all of the allowances within the system for free. Even then, an emitter within the system may have an allocation shortfall and thus may have to bear the cost of buying additional allowances in the market.

In the case where pass-through of the carbon cost is possible, carbon leakage is still a factor to consider. The external supplier into the same market now sees higher prices that may lead to increased profits, again leading to a competitiveness concern.

In January 2015 the EU ETS was 10 years old. There were those who said it wouldn't last and any number of people over the years, who have claimed that it doesn't work, is broken and hasn't delivered. Yet it continues to operate as the bedrock of the EU policy framework to manage carbon dioxide emissions. The simple concept of a finite and declining pool of allowances being allocated, traded and then surrendered as carbon dioxide is emitted has remained. Despite various other issues in its 10-year history the ETS has done this consistently and almost faultlessly year in and year out. The mechanics of the system have never been a problem.

Since the first day of the first draft directive through to present attempts to recalibrate the EU ETS for the period 2021−2030, the overall policy framework within which the system sits has been at the mercy of policy makers, various proponents and opponents of emissions trading and the global economy. Almost none of these have helped deliver a system that actually does what it says on the tin, with the result that the much sought-after carbon price has been largely absent — even as the EU ETS continued to function.

Putting to one side the preliminary 2005–2007 start-up period, the first real carbon price emerged in mid-2007 as power generators started locking in their margin on future electricity sales and became active in the 2008–2012 carbon market. The cost of carbon dioxide emissions was in the range €20–30 per tonne, which was sufficient to see some fuel switching from coal to gas and new emission reduction projects in existing facilities.

Refinery engineers in Pernis, the Shell refinery in Rotterdam, showed me the work they were doing to establish projects across the refinery to tackle emissions. The refinery implemented a project to sell carbon dioxide to greenhouses in Rotterdam, which would displace the burning of natural gas as the source of the gas, used to enhance plant growth by the farmers.

But in 2008 the global financial crisis and EU recession meant a scaling back of EU industrial activity, with a consequent fall in emissions. Not surprisingly, this had an immediate impact on the price of EU allowances. In addition, the inflow of CERs from the Kyoto Protocol CDM was well underway, arguably through a gateway that had been left too widely open as a result of advocacy in earlier years when concerns about high prices were rampant but also from various bodies looking for the EU to support clean development outside its borders. These two factors contributed to a 2 billion-unit allowance surplus that remained for years and held the price at around €5. There was also one significant design flaw in the EU policy framework that added significantly to the oversupply of allowances in the EU ETS.

DESIGNING THE POLICY FRAMEWORK

The phrase 'Potemkin village' was originally used to describe a fake village, built to impress. According to the story, Grigory Potemkin erected fake settlements along the banks of the Dnieper River in order to fool Empress Catherine II during her journey to Crimea in 1787 and reassure her that Russian settlers were arriving. The same phrase is now used, typically in politics and economics, to describe any construction (literal or figurative) built solely to deceive others into thinking that a situation is better than it is. So what have Potemkin villages got to do with the allowance surplus and consequent low allowance price in the EU ETS? The answer, as it turns out, is everything.

The EU ETS should be the cornerstone of the EU's actions to mitigate carbon emissions and as an exemplar for other countries. However, many now perceive that the EU ETS has become more of a compliance formality than an investment driver. This experience demonstrates the negative consequences of overlapping and competing policies and regulations. The fall in price to around €5 throughout 2014 and beyond has gone hand in hand with a growing internal surplus of allowances and of course, no supply side mechanism to stop the flow of new allowances into the system. It was never intended that a huge surplus of allowances should develop in the system; various EU Commission design projections saw a price of some €30 extending into the 2020s. At €5 per tonne of carbon dioxide, the price is not representative of the cost of reducing emissions — it's barely even a holding price.

In effect, this means that the EU doesn't currently have an explicit carbon price to drive change in energy and infrastructure investment — this despite 10 years of policy in place designed with that single goal in mind. The very low price level also implies that there is no expectation of a real carbon price ever developing. In theory these allowances could be bought by a market participant and banked through to Phase IV in the 2020s. Assuming a cost of capital of 5% (and of course availability of capital to do so), a €5 allowance would only need to fetch about €11 in 2030 to cover this, which would be well below the price of a market which is presumably driving investment in carbon capture and storage, surely a technology being seriously considered by then. So what is the thinking that might lead to this discount in market value?

The most plausible explanation is that other policies will be doing the heavy lifting, leaving the ETS as a cosmetic policy instrument. The dominant policies will be on-going renewable energy targets, efficiency requirements and possibly even Emission Performance Standards.

The Renewable Energy Directive has brought projects forward which probably would not have happened until much later in the 2020s. This has had multiple knock-on effects within the EU energy system because of the presence of the ETS and its allowance-based compliance. Whereas the 2020 goal might have been met through improvements in efficiency, fuel switching and the initial phase-in of mature renewable energy technologies (all driven by the cost of carbon dioxide emissions), it has instead been met through a much less cost-effective approach, which forces the implementation of costlier renewable energy projects first. The visible carbon price falls as allowances are freed up with the new investments, but a higher cost remains hidden within the economy. With a low ETS allowance price and with coal use increasing in Europe, when it should be declining, policy makers may turn back to more direct policy instruments to limit the use of coal, which would further undermine the allowance price. This, in combination with yet another round of renewable energy targets and an aggressive energy efficiency drive, exacerbates the situation, leading carbon market traders to take the view that their allowances will have minimal value no matter how long they wait.

The real discussion needs to be around the role of competing or overlapping policies when an emissions trading system is in operation. In fact, the measures now introduced to remove the surplus in the EU may still have insufficient impact if the competing policy issue is not addressed. The end result might be a supply side mechanism that chases an on-going flow of allowances thrown up by other energy system policy measures.

In April 2013 the UK government introduced a UK-specific carbon price floor for power generators. This mechanism operates in tandem with the EU Emissions Trading System, effectively underpinning it for one sector within its existing coverage and therefore encouraging low carbon investment in that sector. The floor is a minimum price of £18 per tonne of carbon dioxide through to 2020, delivered via a top-up payment directly to the government should the EU ETS be trading below £18. If the EU ETS price is high enough, then this policy measure will have no fiscal impact. But in the interim it will serve to support investments in the United Kingdom and deliver a degree of certainty to those making them.

The understandable UK decision to support investment in its power sector doesn't just impact the United Kingdom. In a scenario of lower EU ETS prices, a floor price in the United Kingdom will have an impact right throughout the EU because of the reach of the ETS. Additional UK reductions will almost certainly take place, but an equivalent amount of reductions are now no longer necessary in the EU given that they have been delivered in the United Kingdom, so there is a perfect offset. The overall EU carbon price falls as a result, although the UK electricity consumer does not feel this. The net environmental impact of such a policy is zero because of the role of the EU ETS and allowance trade. Nevertheless, the investment outcome in the United Kingdom is transformed even though the total cost of meeting the EU 2020 target rises as the cost of the extra carbon reductions in the United Kingdom is undoubtedly higher than the next best price on offer in other parts of the EU. The UK consumer bears this additional cost.

While the biggest impact of policy overlay seems to reside in the EU, this is not the only ETS beset by this practice. The California cap-and-trade system started up in early 2012 and is in the process of repeating the EU experience, although going another step further by even having two carbon prices operating in the same sector.

In the road transport sector, which is covered by the cap-and-trade system, the Low Carbon Fuel Standard requires a 10% reduction in the carbon footprint of transport fuels by 2020. This can be achieved through electrification, changes in the well-to-tank emissions of the fuel (e.g. through lowering refinery emissions) and substitution of gasoline with alternatives such as ethanol. The system operates as a baseline-and-credit system would, which effectively delivers a second carbon price into the economy.

For California, as in Europe, investment is increasingly driven by other policies on the back of a specific, predetermined design outcome for the future energy system — almost certainly a higher-cost solution for the energy consumer, but with the same environmental outcome as the cap-and-trade would deliver if left to function on its own. In California this appears to be by design, with the cap-and-trade system operating more as a safety net or safeguard mechanism in case other policies fail to bring down emissions. This isn't really the purpose of a cap-and-trade approach; academic literature argues for it to be the primary driver of change to achieve lowest-cost abatement.

All of which brings the analysis back to Potemkin. The carbon market is providing an apparent low price facade, hiding the real cost of emissions reduction throughout the economy. The policy framework that surrounds the emissions trading system is dampening its effectiveness and denying society the economic efficiency that it would otherwise bring to the task of reducing emissions. Further, by forcing the explicit carbon price to a much lower level with other policies, fuel switching to natural gas is curtailed and the development of CCS is stalled. There is evidence that this is happening in the EU from the observed increase in coal use and the collapse of the NER300 for CCS projects. Even as emissions under the cap fall from the early implementation of renewable energy polices and other measures, the challenge of finding further reductions remains, particularly when the cap covers large swathes of industry such as cement manufacture and steel production. CCS eventually becomes a necessity, yet its introduction and scale-up, which will take some years, is being hampered by the very system that ultimately requires it. Early projects are essential, driven by a meaningful cost for emitting carbon dioxide as the justification for long-term investment.

A cap-and-trade system that is free of policy overlay can deliver the emission reductions that are needed across an economy at lowest cost, but the time span to bring the full range of technologies and mitigation options into the energy system could be forty years, or longer. The EU ETS started in 2005 and should be delivering an 80% reduction in emissions by 2050, a span of forty-five years. But if the real intention of the policy maker is to pre-determine the energy mix or the deployment date for a particular technology, then a cap-and-trade approach becomes largely redundant — however, the cost of implementing the desired emissions outcome potentially soars. This is the real lesson from the EU ETS, but one that has yet to fully hit home in Brussels and other capitals.

SPENDING THE MONEY

Taxes are always an important issue in politics, with political parties often seeking election on the back of their taxation policies and proposals. Given that all carbon pricing systems either raise revenue for the government or shift monetary flows within the economy or both, managing an environmental issue suddenly becomes financially as well as politically important.

Whether it is via the auction of allowances or the taxation of carbon emissions, climate policy is increasingly being seen as a source of revenue for the national treasury. The British Columbia carbon tax of C$30 per tonne of carbon dioxide will raise over $1 billion per annum. EU member state revenues from the ETS have increased as power generators, in particular, must now acquire their allocation of allowances through a government auction, rather than the mainly free allocation that existed between 2005 and 2012 in Phases I and II of the system.

The issue that the collection of revenue raises is what to do with it, how those that pay it are affected and what they get in return. Government already has a long established process for making these choices. Revenue collection is targeted across certain parts of society, money flows into the national treasury, and spending and social welfare provisions are set through the annual Budget. The principal link between revenue collection and spending is the agreement on the size of the deficit or surplus — otherwise the two are largely independent.

But carbon-based revenue has challenged this model. Collection might come from those less able to pay, and many have argued that the money should be used for specific purposes designed to encourage the transition to a near-zero emissions economy. For example, the EU ETS Phase III Directive (for the period 2013–2020, when auction revenues increase significantly) proposes that at least 50% of the revenues generated from the auctioning of allowances be used for emissions management and even presents a list of possibilities.

But why should the revenue from a carbon-pricing system be any different than other revenue that government collects? After all, it is good fiscal practice that revenue collection and spending are two distinct and separate processes. This distinction is also true for large companies, where capital investment on new projects is aligned with future strategy and not decided by the business unit within the company that happened to generate the most cash in a given year.

The decision to shift the energy system to lower emissions and eventually to a near-zero emissions state brings with it a number of questions, the principal one being: *how*? The range of energy technologies required to do the job may not be sufficiently mature or cost-competitive or, worse still, may not exist in a deployable form. This means that technology development must be part of the policy approach required for the energy transition and therefore the government should support energy technology research. In fact, supporting research is an essential element of good climate policy and will mean an increase in government spending. This need for support fosters a natural tendency to link revenue collection and expenditure.

The alternative approach is to delink the collection of revenue and its use, which, as already noted, is the standard practice for most government expenditure. After all, why should the collection of carbon revenue and the fiscal needs of the economy en route to a much lower emission state follow the same trajectory? Today, government expenditure on research and development and direct support to encourage nascent low-carbon technologies (CCS, solar thermal, etc.) may require very significant funding, particularly for large-scale demonstration of these technologies. Problematically, at this stage of price implementation the revenue may be quite low as the government chooses to introduce a new carbon emissions tax at a modest level or to give the bulk of cap-and-trade system allowances away for free to address competitiveness concerns and encourage interest and support by industry groups.

By contrast, looking ahead, carbon revenue may be sizeable and in excess of the transitional needs of new technologies. In this case, forcing the use of a large revenue stream on specific energy system objectives may become a market distortion in itself. It is the job of the underlying mechanism (e.g. carbon tax, cap-and-trade, energy pricing) and the market it creates to drive deployment of a new set of energy technologies, not the job of the government and certainly not the job of the revenue from the pricing mechanism. In the case of a cap-and-trade system, a parallel measure that drives deployment of a particular technology and therefore reduces emissions will lower the carbon price, which in turn impacts the potential for revenue collection. This then argues for the bulk of the money to flow into general revenue that could perhaps lead indirectly to reductions in other taxes. But in doing so, the government must not bypass the need to support technology development, particularly through high-profile demonstrations of key pre-commercial technologies. Perhaps the middle ground here is to have some very specific, time-bound carbon revenue hypothecation against clear objectives (e.g. a large-scale

CCS demonstration), ensuring that this revenue stream doesn't become long-term support for deployment.

Consumers will be out of pocket as a result of the implementation of the pricing mechanism and will look to government for compensation. With additional revenue now in the national accounts, the government can potentially offer compensation for the price increases by lowering taxes. This situation has given rise to the concept of the revenue-neutral carbon tax that has become popular in North America. The Canadian province of British Columbia specifically describes its carbon tax as revenue neutral, and the legislation that introduced it includes specific provisions for recycling the money into the economy, notably as tax reductions or welfare payments. This latter step allows the government to address issues raised by proponents of climate justice who want assurance that those who can least afford the price rises in the economy are suitably compensated. Such compensation measures become problematic if the carbon-pricing revenue is being earmarked exclusively for special emission-reduction purposes.

Another major call on the revenue comes from those who contributed significantly to it — the industrial emitters. Carbon leakage remains a live issue with only partial (but in any case, non-harmonised) implementation of carbon pricing around the world. Free allocation of allowances or taxation rebates can address carbon leakage, but such policies give the government a delicate balancing act to manage. Under-allocation can mean that the competitive disadvantage is not fully addressed, but over-allocation can be classed as state aid, i.e. subsidising particular industries, which may fall foul of rules under international trade agreements.

THE BUSINESS RESPONSE

The transition from an economy that emits carbon dioxide from its energy system to one that doesn't is going to have a cost associated with it and therefore an argument about who is going to pay. After all, if the transition had net benefit and saved everybody money, in a market-based economy, it should be happening anyway. Clearly, that isn't the case.

Energy policy and legislation over the last two decades have shown that many national governments are determined in their efforts to alter the energy mix and to manage emissions, whether for environmental reasons, concerns about energy security or long-term fiscal security. How should business respond to this trend, and what role should it play?

In 2001, the Bush administration proposed a science-and-technology-based approach to reducing emissions, built on further government investment in research and development. Environmental groups, in particular, were not enthusiastic. Even with some business groups as allies, the initiative failed in key areas, such as the development of carbon capture and storage. Had real progress been made, rollout of that technology might have been further along today.

Eight years later a new proposal for reducing emissions came from the Obama administration in the form of a national cap-and-trade system in combination with technology incentives, but Congress failed to approve the legislation. Both the Bush and Obama policy frameworks were relatively good for business — the first because it represented an early start and would have been largely government funded and the second because the overall structure of the approach offered significant competitive protection for key industries and included both a long lead time for implementation and a soft start.

Under the Obama administration, government action to reduce US emissions shifted to the Clean Air Act, but it appears to be the least palatable approach from a business perspective and would likely prove less effective in terms of reducing emissions than the 2009 cap-and-trade proposal, or even a future carbon tax, an option still under consideration by some in the United States. Implementing the provisions of the Clean Air Act could also result in very specific actions on the part of large emitters, meaning that the implied carbon cost for some facilities may be very high. In addition, detailed regulations that address individual sources of emissions may still not result in an overall reduction of national emissions because no overall cap will be in place.

The increasing number of standards-based approaches that are now emerging would seem to highlight the wisdom of establishing a pricing mechanism, but particularly the cap-and-trade approach with all its inherent flexibility. It offers compliance flexibility through offset mechanisms, banking and limited borrowing, competition protection through free allocation in the early phases of implementation, and even technology incentives through constructions such as the NER300 in the EU ETS. Most, if not all of these flexible measures can also be introduced into a carbon tax system, but by way of contrast, a standard has limited flexibility, no price transparency and potentially onerous penalties.

In 2005, despite some initial grumbling, businesses in Europe accepted the proposal of the EU ETS. But its effectiveness has slowly eroded over time. As already noted, this is due to the drop in EU industrial activity from the 2009 recession, the allowed inflow of CERs from the CDM, and the layering of competing policies, which lower the need for allowances and therefore impact the prevailing price. The business community has remained split over what to do about this. Many support the removal of allowances, which Europe has now implemented, but others continue to argue that the system is responding to events and should be left to its own devices. The problem with the latter position is that it could result in an ETS that becomes politically and economically irrelevant because of a sustained low price for allowances that will clearly not drive investment. Lack of effective policies could eventually leave a regulatory-based approach as the way forward in Europe as well as in the United States. Perhaps it is time for the business community to reiterate the case for policies that place a direct cost on carbon dioxide emissions. Economists have argued that case for over two decades, yet policy makers appear to be struggling with implementation. The business community and society more broadly should embrace this approach to dealing with emissions. Here are the top 10 reasons why:

- Action on climate in some form or other is an inconvenient but unavoidable inevitability. Direct standards-based regulation can be difficult to deal with, offer limited flexibility for compliance and may be very costly to implement. The business community is ideally placed to respond to a market price; it does it all the time.

- A direct cost for emitting carbon dioxide, either through taxation or a cap-and-trade system, offers broad compliance flexibility and provides the option for particular facilities to avoid the need for immediate capital investment while still complying with the requirement.

- A carbon price offers technology neutrality. Business and industry are free to choose a path forward in response to the carbon price rather than being forced down a particular route or having market share removed by decree.

- Pricing systems offer the government flexibility to address issues such as cross border competition and carbon leakage (e.g. tax rebates or free allocation of allowances). The EU has a proven track record, with trade-exposed industries receiving a large proportion of their allocation for free.

- A cost for emissions is transparent and can be passed through the supply chain, either up to the resource holder or down to the end user.

- A well-implemented system to deliver a cost for emitting carbon dioxide ensures even economic distribution of the mitigation burden across the economy. This is important and often forgotten. Regulatory approaches are typically opaque when it comes to the cost of implementation, so that the burden on a particular sector may be far greater than initially recognised. A carbon-trading system avoids such distortions by allowing a particular sector to buy allowances instead of taking expensive mitigation actions.

- A carbon price steers the economy towards the lowest-cost pathway for reducing emissions, which also minimises the burden on industry and people in society.

- A carbon price encourages the development of carbon capture and storage, a societal must-have over the longer term if the climate issue is going to be resolved. Further, as the carbon-pricing system brings in new revenue (e.g. through the sale of allowances), governments can use this revenue to support technologies such as CCS.

- A cost associated with emissions of carbon dioxide encourages fuel switching in the power sector, initially from coal to natural gas, but also to zero-carbon alternatives such as wind, solar and nuclear. Encouraging the use of natural gas and other coal alternatives offers a rapid pathway for the economy to reduce emissions. Coal is twice as carbon-intensive as natural gas in power generation.

- It's the smart business-based approach to a tough problem and one that delivers on the environmental objective.

5

THE PARIS AGREEMENT

In March 2015 I was fortunate to visit the Antarctic Peninsula for a second time. Sailing into Neko Harbour and looking past a pod of humpback whales, the edge of the world's largest cache of fresh water stood before us in the form of ice stored in the mighty glaciers of the Antarctic continent. I was there at the invitation of the 2041 organisation talking to a group of young people about climate change and the implications it has for them as the shapers of the 21st century. 2041 has a single mission — preserving the majesty of this mighty continent by leaving it untouched, in part through dealing with the climate issue.

On the day we passed Base Esperanza at the very tip of the Peninsula before turning into the Weddell Sea, the weather was magnificent — so good in fact that the advice we had from our expedition leads as we headed out on every Zodiac excursion, 'Layers, layers, layers (of clothing)', seemed almost superfluous. Two days later Base Esperanza recorded the highest ever temperature in

Antarctica, 17.5°C. A headline in Australia's *Sydney Morning Herald* at the time noted that Antarctica was warmer than Melbourne. A chain of events is now unfolding leading to a gradual reduction in ice mass in Antarctica, thereby raising sea levels and slowly impacting coastlines the world over. I am not one to argue that a single weather event is proof of climate change, but being there and seeing this extraordinary place and feeling the sun on a cloudless day with a new temperature record in the making, it's hard not to think about what is happening and what it means.

However, the participants on my expedition, mainly twenty-somethings, will have to manage the legacy of a higher level of carbon dioxide in the atmosphere that could chart a very different course for Antarctica.

My generation can't correct the problem before we hand over, but we can at least implement the tools the next generation will need to turn the corner and actually start reducing global carbon dioxide emissions. The approaches aren't numerous or complex; however, they do require political and societal willingness to proceed. That willingness initially came to the fore in 1997 in Kyoto, Japan, but the debate it generated lingers on today.

The effective starting point of the global effort to tackle climate change had been the creation of the UNFCCC in 1992, coming on the back of the first Earth Summit, held that year in Rio de Janeiro. By 1997 the parties to the Framework Convention had made astounding progress, agreeing on the Kyoto Protocol and its underpinning trading regime that was designed from the outset to see a cost develop for major emitting economies should carbon dioxide emissions continue to rise. That cost would act as an economic incentive to reduce emissions. The Kyoto Protocol was the very beginning of government-led action to introduce carbon pricing.

By 2001 the EU was responding to the pricing architecture of the Kyoto Protocol and progressing with its design of an emissions trading system (ETS) and other Kyoto-capped countries were beginning to consider similar domestic policy instruments. The move by the EU Commission also introduced the business community into the carbon price discussion in a big way. In fact, I hadn't been in my job as Group Climate Change Adviser for long when a colleague walked into my office and dropped a document onto my desk, asking me to comment on it. That document contained the first draft of the European Union Directive for an Emissions Trading System. But at about the same time President George W. Bush announced that the United States would not ratify the Kyoto Protocol, and that the United States was therefore not bound to its provisions. Instead, he confirmed that the United States would pursue an alternative path on climate change and emissions management.

The Bush announcement was hardly a surprise, given that the US Senate had passed a resolution with a 95-0 vote that the United States should not ratify any agreement that was argued at the time could seriously harm the economy of the United States. Two major issues had surfaced:

- A fairness issue whereby it was argued that as the emissions of many developing countries were rising rapidly and that the day could be seen where Chinese emissions would pass US emissions (that day is now history — the crossover point was in 2006), that more should be asked of these countries under an international framework that was meant to deal with global emissions. A famous chart appeared at the time that showed that even if the Kyoto developed-country emissions dropped to zero between 2000 and 2050, global emissions would keep rising.

- An economic issue, in which it was claimed that the implementation of the Kyoto Protocol would damage the economy of the United States. The cost of reducing emissions and the prospect that such costs would not be shared equitably across the world was becoming problematic.

But there was a third issue, relating to sovereignty as a result of the allocation of allowances by the United Nations to the United States. The implications of a potential loss of national control over the future of the US energy system were too significant to ignore.

Despite the move by the United States the Kyoto Protocol was eventually ratified in 2005 when Russia signed on the dotted line, but the United States had nevertheless triggered the beginning of the end, even before the first compliance period started. By 2007 at COP13 in Bali, a new negotiating track was emerging designed to include the United States, but in so doing the UNFCCC process became increasingly complex and fractured. This eventually led to the collapse of the negotiations in Copenhagen in 2009.

Although the Kyoto Protocol was ratified in order that the first compliance period could start on schedule in

2008 and the 2012 Doha Amendment extended its reach through to 2020, albeit with a reduced number of developed countries, the overall implementation never reached its intended scale and ambition. Canada formally withdrew in late 2011 and compliance was always a formality for the former Soviet states given that their 1990 base-year emissions were taken from their old economy, prior to the economic collapse of the bloc and consequent huge fall in emissions.

The need to manage global emissions and put a halt to the relentless build-up of carbon dioxide in the atmosphere requires the intervention of governments and cooperation between them to ensure their success, particularly when implemented through a cost on carbon dioxide emissions. There is an ongoing debate around the role of government and the scale to which it should be allowed to address the issue of global warming. There are many who believe that government should have only a modest role in society; others accept a much wider role, including one to solve broad-based issues that affect society at large; for example, the build-up of carbon dioxide in the atmosphere. For the latter group, a carbon price may not go far enough; it is a tool designed to tease out the solution over a generation or more. In the case of those who seek to limit the role of government, the imposition of a pricing mechanism across the entire economy can be seen as a step too far and may even raise questions about the foundation upon which the mechanism is based, namely the science of climate change.

Interventions have been attempted at both the national and global level since the atmospheric concentration of carbon dioxide passed 350 ppm more than twenty years ago, a level already well above the pre-industrial concentration of 275 ppm. Yet society knows how to solve the issue of continued carbon dioxide emissions and has known for more than a generation. President George Bush made that clear in his Kyoto exit speech on 11 June 2001 when he noted only two ways to stabilise the concentration of greenhouse gases in the atmosphere; these are to avoid emitting in the first place or to capture afterwards.

Somehow the route forward has been complicated to such an extent that resolution seems more distant than ever. The solution framework offered by the Kyoto Protocol has effectively ended, but there remains inevitability around political action and legislation to deal with carbon dioxide emissions. It's not just politics that dictates this, but physics as well. Society can't keep on adding carbon dioxide to the atmosphere and expect nothing to change.

In countries with strong social underpinnings, political action might happen quite quickly, while in others there will be debates for years or perhaps largely ineffective measures will win the day. In other countries, the task is still too daunting to undertake, particularly in economies that are predominantly tied to fossil-fuel use or lack the political stability necessary to embrace change.

In my professional occupation, I have witnessed each of these scenarios. The European Union moved ahead quite quickly with the implementation of its emissions trading system. South Africa was quick to recognise the problems associated with its significant use of coal, but is challenged by a willingness to act on this in the face of an economy that needs energy for its people and sustained growth to raise living standards. Australia, my home country, has found the national debate to be tough going. It is a resource economy with high per capita emissions and vast fossil energy reserves. After a decade of trying to reach a consensus on resolving the issues regarding climate change, a system to push the country towards its 2009 Copenhagen emissions reduction pledge did not fully emerge until 2014. But even that system is faltering.[29]

By contrast, Canada has now found a very different way forward, even though it narrowly missed take-off. Under the Kyoto Protocol, Canada ratified a national target that was more ambitious than those put forward by economically comparable countries. Given that its electricity system was already very dependent on hydro and, to a lesser extent, nuclear, setting such a target was quite admirable. Shortly after Canada ratified the Kyoto Protocol, I had a conversation with a senior Canadian politician that went something like this:

CP (Canadian politician): The target is very difficult to meet.

Me: Agreed, but perhaps that shouldn't be your primary consideration.

CP: It has to be; we have a target.

Me: Yes, but perhaps you should focus on getting a carbon price embedded in the economy first, then use that to start to drive change.

CP: But will we meet the target?

Me: You may not, but you would leave a legacy of an economy with emissions management up and running, the required capacity building done and emissions at least moving in the right direction. At the end of the day if you don't meet the target, at least you will have made a good attempt.

CP: *Yes, I understand. But what about the target?*

To this day the Canadian federal government has struggled with adopting an appropriate national implementation plan to reduce emissions. Early target paralysis on the back of an ambitious pledge may well have been a contributing factor to its plight.

GOING LOCAL

Over the intervening years, action has devolved to the provinces and real progress has been made, with British Columbia having one of the most robust and comprehensive carbon taxation systems running in the world today; there is also strong evidence that it is working. By contrast, Québec has joined the California emissions trading system through international linkage and Ontario is following. Canada has also emerged as a global leader in CCS through provincial actions in Alberta and Saskatchewan. Both by design and by default the Canadian provinces have got on with the job of managing emissions in ways that suit their own economies. This may not be as robust as a single national approach, but perhaps more importantly, it is politically achievable.

At the international level the economic clarity delivered by the Kyoto Protocol and its attempt at global carbon pricing through a single approach is being replaced by a series of actions implemented by decision makers at local, national and regional level. This switch from global to local has now become the foundation for the Paris Agreement.

But this wasn't the only difference between the near collapse of the UNFCCC process at COP15 in Copenhagen where the notion of a single global approach persisted and COP21 in Paris, where a successful outcome prevailed. On almost every front the differences between 2009 and 2015 were notable.

2009	2015
Months before COP15, 'climategate' saw an unknown group hack into the University of East Anglia e-mail system and publish thousands of documents. It was then claimed that the documents showed evidence of fabrication of the warming trend through temperature data manipulation. No such conspiracy was ever proven.	When the monster 2015 El Niño settled in across the Pacific and global temperatures rose to record levels, there was an almost discernible sigh of relief from the climate science community. The arguments of a pause in warming were swept away as the 2015 temperature topped out at almost $1.1°C$[1] above pre-industrial levels.[2]
Politically, the big questions were still being debated: Would the agreement be top down like the Kyoto Protocol with national targets established through the UNFCCC? Or would it be bottom up from self-determined national actions? And what would happen to the Kyoto Protocol? Should there be a global goal to limit temperature rise? How should the developing countries be treated?	Although there were myriad details to be agreed upon, the Paris negotiators were in agreement on what it was they were trying to agree on: a relatively lean framework within which could sit the collection of nationally determined contributions (NDCs) from all countries for scrutiny and review. It had taken decades to reach this point, but the task at hand was clear.
All eyes were on the deliberations of the US House of Representatives and the Waxman-Markey cap-and-trade bill, with every expectation that the United States would take the lead and establish an economy-wide carbon price. This didn't happen.	Eyes are now on the world's largest emitter, China, as it proceeds with its carbon-pricing provincial trials and expansion to a nationwide system. By contrast, the United States has started to implement a command-and-control regulation-based approach in the power sector.

(Continued)

2009	2015
At the June 2009 UNFCCC meeting, just six months prior to Copenhagen, the team from the Oxford University Department of Physics first presented their new thinking on a global carbon emissions limit of 1 trillion tonnes over the industrial era.	*Negotiators have embodied the concept of net-zero emissions within the Paris Agreement and therefore an end date to the on-going accumulation of carbon dioxide in the atmosphere.*
The British government produced a first-of-its-kind report on the idea of global carbon trading.	*The World Bank is now taking the concept forward. A linked market exists between California and Québec.*
I came across the first EV charging stations in London and met a person who was taking delivery of the seventh Tesla in the United Kingdom	*There were approximately 28,000 EV and PHEV newly registered vehicles in the United Kingdom. Sales jumped 94% compared to the previous year and there was a market for used Teslas.*
Within the UNFCCC process, negotiations were operating on two tracks, the Kyoto Protocol and Long-term Cooperative Action. There was little consensus on climate finance and almost no sign of targets and goal setting from the major developing countries.	*Some 190 countries submitted NDCs prior to the start of COP21. China announced a plateau in national emissions by around 2030 and other countries followed its lead.*

[1]NOAA.
[2]1850–1870 average, HADCrut4 series.

DESIGN CONSIDERATIONS

But even as the shape of the Paris Agreement was becoming clear, a barrage of suggested alternative routes forward continued to appear. This was part of a salvo of pre-Paris articles that sought to direct the negotiations towards a particular solution space, including many that presented the case for carbon-pricing systems. Two articles from *Foreign Affairs* in mid-2015 offer good examples of this practice: one, by Michael Bloomberg, argued that the emissions-mitigation solution increasingly lies with cities and the other put the challenge squarely in front of the business community. Neither approach would be sufficiently comprehensive for an issue as deeply rooted as the global energy system dependency on fossil fuels.

This multitude of opinions and possible solutions were reminiscent of 2010's Hartwell Paper[30] published by the London School of Economics in conjunction with the University of Oxford after the failure of Copenhagen. In it a group of UK economists cast the climate issue as a 'wicked problem'. This is a type of problem that is difficult or impossible to solve because of incomplete, contradictory and changing requirements that are often difficult to recognise. The use of the term 'wicked' here has come to denote resistance to resolution, rather than evil. Moreover, because of complex inter-dependencies, the effort to solve one aspect of a wicked problem may reveal or create other problems.

Approaching a 'wicked' challenge raises questions about which problem is actually being solved. It may turn out that the issue of climate change is immensely more difficult to solve than the issue of carbon dioxide emissions. There is now good evidence that bringing emissions down to near-zero levels doesn't necessarily lead to immediate resolution of the changing climate.

Increasing levels of carbon dioxide in the atmosphere is driving current warming of the climate system, but the scale on which anthropogenic activities are now conducted can impact the climate through different routes.

Moving away from fossil fuels to very large-scale production of energy through other means is a good illustration
of this. In a 2010 report, MIT showed that very large-
scale wind farms could result in surface warming because
the turbulent transfer of heat from the surface to the
higher layers is reduced as a result of reduced surface
kinetic energy (the wind) that is converted to electricity.
This is not to argue that wind turbines shouldn't be built,
but rather to highlight that with a population of 7–10
billion people, society may inadvertently and unwittingly
engage in some degree of geo-engineering (large-scale
manipulation of an environmental process that affects the
earth's climate) to supply it with goods and services.

Even narrowing the broader climate issue to emissions,
the problem remains wicked. Inter-dependencies abound,
such as when significant volumes of liquid fuels are supplied by very large-scale use of biomass or when efficiency
drives an increase in energy use (as it has done for over
100 years), rather than the desired reduction in emissions.
Even the desired energy transition itself may produce a
significant emissions bulge as a massive new infrastructure build is undertaken.

Any approach to managing wicked problems involves
carefully defining the problem first. This might involve
locking down the problem definition or developing a
description of a related problem that you can solve, and
declaring that to be the problem. Objective metrics by
which to measure the solution's success are also very
important. In the field of climate change and the attempts
by the Parties to the UNFCCC to resolve it, this is far
from the course currently being taken. There is immense
pressure to engage in sustainable development, end

poverty, improve access to energy, promote renewable technologies, save forests, solve global equity issues and use energy more efficiently. Although these are all important goals, they are not sufficiently succinct and defined to enable a clear pathway to resolution, nor does solving them necessarily lead to restoration of a stable climate. The nationally determined contribution-based approach allows for almost any problem to be solved, so long as it can be loosely linked to the broad categories of mitigation and adaptation. So is the global approach agreed upon in Paris adding to the confusion rather than simplifying or even avoiding it?

A sobering observation from a deeper analysis of wicked problems is that attempting to solve such a problem may fail in the long run, even when the short-term solution appears appealing and well suited. Due to its complexity, the problem may simply reassert itself, perhaps in a different guise — or worse, the solution could exacerbate the problem.

In climate-change terms, this translates to emissions not falling as a result of efforts made, or even if they did fall somewhat, the fall having no measurable impact on the continuing rise in atmospheric carbon dioxide levels.

That is not to say that society should give up. As the counter to this observation, having defined a clear and related objective to the wicked problem, there are a few possible solutions, and the focus must be on selecting from amongst them. If the objective is to reduce emissions or reduce the overall use of fossil fuels, solving the problem comes down to implementing a cost for emitting carbon dioxide (through systems such as cap-and-trade or carbon taxation) and developing CCS. But for the negotiators, different priorities presented themselves, and they set about resolving them well before the two-week marathon that gave birth to the Paris Agreement.

In the years between Copenhagen and Paris, two key issues had to be addressed before any real hope of a global deal could emerge. Both related to the nature of actions to be taken to mitigate emissions.

Resolution of the first issue would require a challenge to the respective roles of developed and developing countries and their responsibilities in relation to emissions mitigation. Following the agreement of the Framework Convention in 1992, these roles were further clarified within the 1995 Berlin Mandate, which set in motion the negotiations that led to the 1997 Kyoto Protocol.

The Berlin Mandate included the following points:

- The legitimate needs of the developing countries for the achievement of sustained economic growth and the eradication of poverty;

- The fact that the largest share of historical and current (in 1995) global emissions of greenhouse gases has originated in developed countries;

- For developed country/other parties included in Annex I, to set quantified limitation and reduction objectives within specified time frames, such as 2005, 2010 and 2020;

- Not introduce any new commitments for Parties not included in Annex I.

While the mandate led to the successful negotiation, agreement and ratification of the Kyoto Protocol, it also led to its eventual demise. In the following two decades the emissions from China in particular, but also other countries, skyrocketed. Therefore, the issue of responsibility for reducing emissions and where the onus should fall became acute. Clearly, developed countries had to take a leadership role in reducing emissions, but the math behind a global reduction had shifted since the 1990s, to the extent that it was no longer possible to see a global reduction in emissions even if the developed countries rapidly headed to zero.

In 1990 the global energy system emissions split was roughly 14 Gt from developed countries (UNFCCC Annex 1 countries) and 6 Gt from developing countries (UNFCCC non-Annex 1). In 2015 the developed countries were still emitting around 14 Gt, but emissions from developing countries had risen sharply to about 20 Gt. This means that even if the developed countries had trended to zero emissions over this period, global emissions would not have fallen. Looking ahead as India and Africa develop rapidly and Asia continues the development seen over the last 20 years, developing countries will have to join developed countries in embarking on a reduction strategy if global emissions are to fall.

This challenge was partially addressed at COP 17 in Durban in 2011 with the Durban Mandate, which broadened the basis for negotiation on emission reductions beyond the original emphasis on developed countries taking the lead. Paragraph 7 of the Establishment of an Ad-Hoc Working Group on the Durban Platform for Enhanced Action is as follows:

> *The Conference of the Parties,*
>
> *Decides to launch a workplan on enhancing mitigation ambition to identify and to explore options*

*for a range of actions that can close the ambition
gap with a view to ensuring the highest possible
mitigation efforts by all Parties.*

According to Professor Robert Stavins, the Director of
the Harvard Project on Climate Agreements, the Durban
Platform represented a dramatic departure from some 17
years of UN-hosted international negotiations on climate
change — the 17th Conference of the Parties in Durban had
turned away from the Annex I/non-Annex I distinction.

The second issue to be resolved was of a more sovereign nature and pertained, in particular, to the United
States. Did a group of nations have the right to insist that
a particular reduction goal should be adopted within the
law of another nation and then strictly adhered to?
Should the answer be in the affirmative, then any future
global deal would require formal ratification by the US
Senate. Such approval is far from a given, even if a particular Administration agrees to the deal in principle. A
global deal could still proceed without the United States,
as it did with the Kyoto Protocol, but the outcome isn't
sustainable, as was also seen with the Kyoto Protocol.

This issue was finally resolved at the Warsaw COP in
November 2013. The agreed Warsaw text changed the
nature and legality of mitigation actions by renaming
them. Specifically, the text says:

*To invite all Parties to initiate or intensify domes-
tic preparations for their intended nationally
determined contributions, without prejudice to
the legal nature of the contributions, in the context of adopting a protocol, another legal instrument or an agreed outcome with legal force under*

the Convention applicable to all Parties towards achieving the objective of the Convention as set out in its Article 2 ...

Reaffirming the mandate agreed in Durban, which aims to see all countries treating mitigation similarly, the negotiators landed on a common requirement to prepare contributions, rather than some countries being asked for specific reduction targets or commitments and others for appropriate (to their development status) actions. The latter would have been a retreat back towards the strict developed/developing country division of the Kyoto Protocol.

But the compromise came with the use of the word contribution, which is not the same as a commitment. The two words have very different meanings:

Commitment: *the state or quality of being dedicated to a cause, activity, etc., or, an engagement or obligation that restricts freedom of action.*

Contribution: *a gift or payment to a common fund or collection (e.g. the part played by a person or thing in bringing about a result or helping something to advance).*

What the world has settled on is essentially a voluntary role for the Parties to the Agreement as this is the essence of a contribution, rather than an obligation. While this neatly sidesteps the sovereignty issue, it is nevertheless the case that nations will have no legal requirement to implement emission reductions. Only peer pressure can ensure that contributions are effectively delivered.

With the twin issues of responsibility and response largely resolved, at least from a diplomacy perspective, the negotiators moved onto Paris with a view to finalising the new global deal.

Some 150 heads of government and heads of state had turned up in Paris to kick off proceedings and although most departed immediately afterwards to leave the job with their negotiating teams, the telephones ran hot between capital cities across the world over the ensuing two weeks. Indeed, it was even rumoured that the newly forged friendship between the United States and Cuba meant that the two countries cooperated to put pressure on Nicaragua when it appeared that its negotiator was going to hold up proceedings with some fiery rhetoric in the final stages of the main plenary meeting.

In the previous eighteen months, staff in French embassies all over the world had worked tirelessly to support the process, but in the end it was the negotiators themselves working through the night in the final days who delivered the deal. All manner of behind-the-scenes trade-offs were made to resolve profound disagreements on a dozen or so key issues, including the temperature goal itself, the eventual need for net-zero emissions of greenhouse gases and the level of financial assistance for developing countries. There were also hundreds of smaller issues and points of principle that got dealt with during the final days, ranging from continued specific recognition of developing countries in certain instances to the role of a non-market mechanism to support mitigation.

In the days before the start of the COP the text had extended to nearly one hundred pages, with multiple variations of almost every clause and hundreds of square bracketed words and phrases, indicating disagreement amongst the Parties. But unlike 2009's COP15, where almost everything that could go wrong with the process ultimately did, the French Foreign Ministry had left nothing to chance and were to be congratulated on an extraordinary outcome. They had marshalled their considerable diplomatic corps throughout the preceding year, staging activities in a series of capital cities, conducting high-level dialogues with business leaders and rallying French industry to act as leaders in global trade bodies. In addition, seminars, conferences and panel events were held throughout the world to build support.

In 2014 representatives of a cross-section of business interests and companies were invited to the preceding New York Climate Summit, where there would be a special focus on carbon pricing that required the support of the business community. I arrived to find a vast demonstration stretching all the way across mid-town Manhattan; it took well over two hours to pass my viewing point and included almost every climate-focused issue group imaginable, along with countless concerned individuals. It was clear, from that moment on, that COP21 would be something special. I left New York with a sense of optimism about the year ahead, buoyed by the fact that carbon pricing was back on the global agenda. But I was enough of a realist to know that the road ahead would be tough. And it was.

City mayors turned out in Paris in force, which added to the momentum. Anne Hidalgo, Mayor of the City of Paris and Michael R. Bloomberg, the UN Secretary-General's Special Envoy for Cities and Climate Change — in partnership with the global networks of cities and local governments for climate action — co-hosted the Climate Summit for Local Leaders — the largest global convening of mayors, governors and local leaders focused on climate change. Over a thousand officials turned up, including Boris Johnson from my home city of London.

COP21 was not entirely smooth sailing though. With the discussions stalling by that first weekend, marking the COP's halfway point, the French implemented a negotiating technique borrowed from COP 17 in Durban to reach a consensus. The negotiating procedure to break the deadlocks that were grinding the process to a standstill took immense skill and perseverance. This negotiating tactic came from the Zulu and Xhosa people and is called an 'indaba' (pronounced IN-DAR-BAH). It involves small groups gathering to focus discussions between the bigger parties. An indaba is designed to allow every party to voice its opinion, and therefore arrive at a consensus more quickly because no one is overlooked.

The process orchestrated by the French government worked. In the end the Paris Agreement came in at under 20 pages and consisted of 29 Articles, supported by some twenty pages of decision text that outlines how the deal should be implemented. It has the potential to solve the climate issue within this century, provided it is implemented with the same focus and attention to detail that the French brought to the negotiation table.

The Agreement goes a long way towards the EU and US goal of having little to no differentiation between countries; in other words, all nations must now take substantial action to manage emissions, irrespective of their development status. At the same time the Paris text clearly preserves the developed/developing country split in a number of areas, but particularly with regard to financial flows, thereby addressing the thorny issue of money for the least developed economies. The Agreement also strikes a delicate balance between being legally binding in its intent and not binding to the extent that mitigation actions can be dictated from above, another important requirement for the United States in particular.

Although a global goal to limit warming of the climate system to 2°C was expected, the surprise outcome of the two weeks in Paris was the formation of a coalition of ambition, which moved the goalposts and set them well below 2°C, with a view to moving them again to 1.5°C. Arguably, this puts the global limit somewhere in the territory between 1.5°C and 1.8°C of warming relative to a pre-industrial baseline. This shift had its foundation in the deep concern of the Alliance of Small Island States (AOSIS) — a group of over forty low-lying islands around the world currently chaired by the Maldives — regarding future sea-level rise. Some are already threatened by rising seas, but the longer-term rise, which could even extend to several metres if the West Antarctic ice shelf begins to seriously deteriorate, would see many of their nations vanish completely. By targeting a lower-temperature outcome, they hope to shift the sea-level rise risk profile enough to significantly lower the chance of such an outcome.

THE END RESULT

The Agreement rests on the concept that nations will bring their own plans or goals for reducing emissions to the table, rather than being set specific goals by the COP. This is the most fundamental change to emerge from Paris and the process that preceded COP21. The Agreement is now rooted in national policy and actions as the determining factor, rather than in the COP or the UNFCCC driving national ambition on a top-down basis.

In exchange for an approach based on self-determination, or contributions (the NDCs), the Agreement specifies a high level of transparency and openness in reporting on progress, reviewing results and attempting to push the goals and plans further on a five-year cycle starting in 2018. From a policy implementation perspective, such a cycle time translates into almost continuous scrutiny and update but particularly in the short term given the gap that must be filled between the current global emissions trajectory and one that is pointing towards an outcome of well below 2°C. Even in the EU, where a 40% GHG emission reduction target (vs. a 1990 baseline) for 2030 has been agreed, there is now a case to improve on this and eventually announce a more stringent 2030 goal.

The Agreement isn't just about reducing emissions. Adapting to the impacts of climate change also features in the text. A global goal is established for enhancing adaptive capacity, strengthening resilience and reducing vulnerability to climate change. The goal is aligned with the new temperature goal and seeks to ensure adequate adaptive response for the particular level of warming. International cooperation with regards implementation of this goal is stressed.

At the Doha COP in 2012 the issue of how to ascertain the means by which reparations for loss and damage associated with the impacts of climate change could be assessed became contentious. Some less developed countries were looking to wealthier high-emitting nations for reparations in relation to climate-induced damage. The discussion around this issue has remained active, but the Paris Agreement has put to one side the possibility of financial claims relating to loss and damage. Rather, the text builds on the agreement reached at the Warsaw COP in 2013 and establishes a mechanism to address loss and damage through enhanced understanding, action and support on a cooperative and facilitative basis.

Article 9 of the Paris Agreement covers the key issue of financial assistance for developing countries. It calls on developed country Parties to provide financial resources to assist developing country Parties with respect to both mitigation and adaptation in continuation of the former's existing obligations under the Convention. Other Parties are encouraged to provide, or continue to provide, such support voluntarily. This section of the Agreement retains a portion of the historical differentiation of responsibilities between developed and developing countries, in contrast to the rest of the text which, for the most part, seeks to remove it. But there was no attempt to recast the respective lists of developed and developing countries that exist under the Convention and are now over twenty years old. The decisions that support the Paris Agreement now set $100 billion per annum as the floor for financial transfers from developed to developing countries, with many developing country NDCs having substantial price tags built into them.

Technology plays a supporting role in delivering on the goals of the Agreement. The existing and long-standing focus on technology that operates under the Framework Convention will continue, but within a broader structure that will seek to expand the development and deployment of key technologies into developing countries with suitable financial support.

Finally, and importantly, what happened to the carbon-pricing provisions that were built into the Kyoto Protocol and are so critical to long-term success? There is a specific mention of the need for carbon pricing in the decision text, but the phrase does not appear in the actual Agreement. However, Article 6 lays out a framework for the transfer of mitigation actions across national borders and the development of a supporting mechanism, both of which will ultimately require market structure and price discovery to function. These are the building blocks of global carbon pricing and emissions trading, although government implementation of national carbon-pricing mechanisms and carbon-emissions accounting is an essential prerequisite. The task now rests with the negotiators to develop a thoughtful and effective implementation of Article 6 to ensure its success as a means towards carbon pricing operating at a global level.

At their first official bilateral meeting in March 2016, President Obama and Canadian prime minister Justin Trudeau made the significant move of responding to Article 6 of the Paris Agreement. They pledged that their countries would work together to support robust implementation of the carbon markets-related provisions of the Paris Agreement, including the transfer of units between respective NDCs. In fact, Canada and the United States are already heading down this pathway with the link that now exists between the Québec and California cap-and-trade systems. Ontario is following the lead of Québec, as might some US states eventually follow California.

As a veteran of some 15 COPs and many inter-sessional meetings, I think it's safe to say that the Paris Agreement is a remarkable breakthrough achievement. But because of its brevity it still needs considerable deciphering and an understanding of the energy system in place today to put it into perspective.

THE STRETCH GOAL

Against a backdrop of continued reliance on coal in particular, the agreement to limit warming to well below 2°C has established a far more ambitious global goal than many imagined possible. The idea of a global goal goes back to the creation of the UNFCCC. In 1992 the original text agreed on the need to stabilise greenhouse gas concentrations in the atmosphere at a level that would prevent dangerous anthropogenic interference with the climate system. At COP16 in Cancun, the Parties to the UNFCCC reformulated this as a numerical goal — the need to limit warming of the climate system to no more than 2°C above the pre-industrial level with consideration for reducing this to 1.5°C as the science might dictate. This seems very clear, but in fact offers little immediate guidance to those attempting to establish a national or even global emissions pathway.

In line with the Cancun Agreements, a number of parties have maintained the need to lower the goal to 1.5°C, but particularly those from low-lying small island states who are justifiably concerned about long-term sea-level rise. Science is certainly on their side. The IPCC report states that the sea-level rise will continue for many centuries beyond 2100, with the amount of rise dependent on future emissions. This could be less than one metre by 2300 for a more modest temperature increase such as 1.5°C but ranging up to three metres or more as atmospheric concentrations of carbon dioxide head to levels well above 700 ppm. Even more significant sea-level rise could result from sustained mass loss by ice sheets, and some part of the mass loss might be irreversible. There is fear that sustained warming greater than a certain threshold above pre-industrial levels would lead to the near-complete loss of the Greenland ice sheet. Although that threshold is estimated to be as little as 1°C, the more likely level is well within the plausible range of temperature rise for this century if emissions continue unabated.

Including a stretch goal to limit warming to 1.5°C was voiced loudly in Paris and gained significant momentum through a growing coalition of countries supported by a strong dose of optimism. With such enthusiasm and following the precedent created in Copenhagen, the wording was included. But the climate system is a slow lumbering beast and the global temperature could take years to settle down once there is stabilisation of carbon dioxide (and other greenhouse gases) in the atmosphere. It could be a decade or more after that before there is certainty that no further temperature rises are in store and that a sustained decline may be in place as the biosphere takes up carbon dioxide and levels in the atmosphere begin to fall.

NET-ZERO EMISSIONS

More recently and because of the cumulative nature of the carbon dioxide problem the concept of net-zero emissions has entered the popular climate vernacular, largely replacing the language of the pre-Copenhagen era, which simply argued for a sharp reduction in emissions by 2050. Net-zero emissions is the point in time at which there is no net flow of anthropogenic carbon dioxide into the atmosphere, either because there are no emissions or, as is more likely, because when emissions remain, they are offset with a set of practices known as negative emissions technologies (NET). A NET is a technology that draws down on atmospheric carbon dioxide; perhaps the simplest implementation of this is planting a tree.

The IPCC 5th Assessment Report notes that as cumulative emissions of carbon dioxide largely determine global mean surface warming by the late 21st century and beyond, this would imply a limit for cumulative emissions of carbon dioxide and that such a limit would require that global net emissions of carbon dioxide eventually decrease to zero. A further look at the IPCC report shows that many of the scenarios they report on that are consistent with the 2°C goal also include a period in the second half of the century when global emissions are actually negative. The reason for needing such a period is that under these scenarios it doesn't prove possible to limit emissions sufficiently, given the time it takes to re-engineer the energy system in the face of rising demand and legacy infrastructure.

Net-zero emissions is important because it represents the point at which there is no further rise in atmospheric carbon dioxide. As such, the need for net-zero emissions has been closely linked to 2°C, but in fact any temperature plateau, be it 1.5°C or even 4°C, requires net-zero emissions. If not, warming continues as atmospheric carbon dioxide levels rise. There is now a discussion as to when net-zero emissions should be reached — as early as 2050 (but practicality must be a consideration), or perhaps by the end of the century. However, the key determinant for temperature is cumulative emissions, which is the area under the emissions curve before net-zero emissions is achieved, less the area under the curve after it is reached. This assumes that emissions eventually trend into negative territory with technologies such as direct air removal of carbon dioxide with capture and storage (DACCS) or bioenergy conversion with CCS (BECCS); both technologies that are now being viewed as critical for long-term atmospheric carbon dioxide management.

DACCS requires the removal of carbon dioxide from the air itself after which the carbon dioxide is stored in geological formations deep below the surface. Some people have imagined vast stretches of artificial trees lining our freeways capturing carbon dioxide akin to natural forests. The reality is more likely to be very large capture facilities that look similar to a modern refinery or chemical plant, but are situated close to a suitable deep geological storage reservoir. The artificial tree model would require too much carbon dioxide handling and transport. Air capture of carbon dioxide is operating on a tiny scale today in a few test facilities where the search is on for a low cost mechanism to separate carbon dioxide from the air.

BECCS is an entirely different approach to air capture. When a plant is grown, it absorbs carbon dioxide from the atmosphere, but in doing so, it converts the carbon dioxide via photosynthesis into complex carbon-based molecules, which make up the structure of the plant, e.g. wood. Wood can be burned in a power station to generate heat and make electricity, which, in turn, releases carbon dioxide. If this carbon dioxide is captured and stored rather than emitted back to the atmosphere, the whole process becomes carbon negative; in other words, it results in a net draw of carbon dioxide from the atmosphere. The production of biofuels may also release carbon dioxide, as, for example, when sugar is fermented in the manufacture of ethanol. This pure carbon dioxide can be captured relatively easily and when stored, introduces a negative emissions benefit to the production process.

An article in *Nature Climate*[31] at the time of COP21 looks more closely at the set of NETs that society may come to depend on. As well as DACCS and BECCS, several others were included:

- Mineral carbonation involves reactions of magnesium or calcium oxides (typically contained in mineral silicates and industrial wastes) with carbon dioxide to give inert carbonates. These reactions occur slowly in nature and over time trap vast quantities of carbon, but pilot facilities have been built to do this on an industrial scale and produce useful products.

- Afforestation and reforestation to fix atmospheric carbon in biomass and soils. Large-scale tree planting increases natural storage of carbon in biomass and forest floor soil. During the Great Depression, the United States planted several billion trees to create hundreds of new national forests. Today, annual net sequestration of carbon in managed US forests offsets approximately 15% of the annual emissions of carbon that result from the combustion of fossil fuels. Further to this, reintroducing wood into housing construction can sequester carbon for decades or even centuries.

- Manipulation of carbon uptake by the ocean, either biologically or chemically. The ocean is a huge store of carbon dioxide, although increasing levels dissolved in the water are raising ocean acidity. However, uptake could be safely enhanced by increasing marine photosynthesis, such as through large-scale cultivation of seaweed in shallow areas.

- Altered agricultural practices, such as increased carbon storage in farm soil.

- Converting biomass to recalcitrant bio char, for use as a soil amendment.

The article focuses on BECCS, DACCS, mineral carbonation and forestry and gives a detailed breakdown of the global impacts of these technology areas in terms of water, energy needs and land use. It is clear that there is no silver bullet to rely on. While BECCS and DACCS can potentially be deployed at scale and make a material difference to atmospheric carbon dioxide (>3 Gt carbon per annum by 2100, or 10+ Gt carbon dioxide), BECCS requires significant land and water use (but is a net energy producer) and DACCS is a big energy user. The latter is also deemed to be very expensive to implement. Mineral carbonation, on the other hand, doesn't make the grade in terms of scale. That leaves afforestation and reforestation, which is certainly scalable but only extraordinary deployment occupying huge swathes of land will make a significant difference in atmospheric carbon dioxide.

The date at which net-zero emissions is reached is important, but it is not necessarily an indicator of the eventual rise in temperature. Just to complicate matters further, although the world needs to achieve net-zero emissions, net anthropogenic emissions do not necessarily have to be zero by 2050 or 2100 to meet a given temperature goal because of carbon removal arising from natural sinks in the oceans and terrestrial ecosystems.

Another form of emission reduction goal is to call for an urgent peaking of emissions. This is important, but again it doesn't tell the full story. What is critical is what happens after the emissions have peaked. A long slow decline to achieve a plateau would be good news, but unless that plateau is close to net-zero emissions, then the cumulative emissions continue to build, along with the associated warming. Still other proposals continue to argue for emissions to reach a considerably reduced level by some future date, which presumes a follow-on trajectory equating to the temperature goal.

The long-term goal embedded in the Paris Agreement was defined not only in terms of temperature but also as a pathway, outlined in Article 2 and Article 4:

Within Article 2:

> *Holding the increase in the global average temperature to well below 2°C above pre-industrial levels and to pursue efforts to limit the temperature increase to 1.5°C above pre-industrial levels, recognising that this would significantly reduce the risks and impacts of climate change.*

Within Article 4:

> *In order to achieve the long-term temperature goal set out in Article 2, Parties aim to reach*

*global peaking of greenhouse gas emissions as
soon as possible, recognizing that peaking will
take longer for developing country Parties,
and to undertake rapid reductions thereafter in
accordance with best available science, so as to
achieve a balance between anthropogenic
emissions by sources and removals by sinks
of greenhouse gases in the second half of this
century ......*

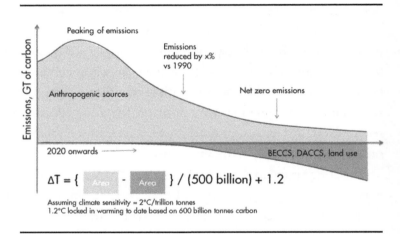

Based on the work of Myles Allen and his colleagues,
the Paris Agreement therefore sets out our need to limit
the cumulative release of carbon to well below one trillion tonnes. But so fast is the current rise that a simple
projection of current annual emissions with some growth
over time shows one trillion tonnes being exceeded before
the year 2040. This suggests the need for drastic change
in the energy system in a very short period to stay within
one trillion tonnes.

There are always a variety of trajectories possible for any temperature goal, but 1.5°C offers little room for flexibility, given its stringency. It can be equated to a cumulative carbon emissions threshold of some 750 billion tonnes, compared with cumulative emissions by the time of the Paris Agreement of around 600 billion tonnes. In such a pathway, global net-zero emissions must be reached in just 40 years, although this still results in some 860 billion tonnes accumulation. Therefore, this must be followed by another half century of atmospheric carbon removal and storage (~100 billion tonnes removal). Some 10 billion tonnes of carbon dioxide must be removed and stored each year by late in the century, possibly using the aforementioned DACCS and BECCS technologies. Significant reforestation would also play a major role. With infrastructure in place, the 22nd century might even offer the possibility of drawing down on carbon dioxide below a level that corresponds with 1.5°C.

The later net-zero emissions is reached, the greater the post net-zero dependence on CCS is likely to become. The pathway outlined above (and almost any other 1.5°C pathway) is completely dependent on it.

Re-engineering the global economy such that it transitions from carbon dioxide emissions of some 40 billion tonnes per annum to a net-zero state in less than one person's working life is an unlikely proposition — however, the maths of 1.5°C demands this or at least something close to it.

Yet some plausible scenarios do reach net-zero emissions during this century, albeit not soon enough to deliver a 2°C outcome, let alone 1.5°C. In 2016 climate scientists and economists at the MITJP used their

Integrated Global System Modelling framework (IGSM) to look at, amongst others, the *Shell New Lens Scenarios*, first published in 2013.[32] The MIT analysis[33] established that the two scenarios developed by Shell offer trajectories of 2.4° and 2.7°C respectively (but within a wide uncertainty band as with all MIT climate projections). They both achieve net-zero emissions around the end of the century (in contrast to ~2050 for a 1.5 + °C goal) through very substantial deployment of CCS, which needs to start in earnest as early as the 2020's. Both require a level of governmental policy action and commitment currently not in place. One requires carbon pricing of $10-$30 covering much of the global economy by 2020 (but with EU at $65) and $25−$90 by the 2030s, with coverage exceeding 80% of economic activity. By contrast in 2015, carbon-pricing policies covered a small fraction of global emissions, with most approaches operating in the range $1−$15 and a few up to $30. However, with the arrival of an economy-wide system in China during 2017, this fraction will jump significantly.

While the case for 1.5°C has certainly been made from a climate perspective, there is no secure plan in place for implementation. Large-scale use of technologies to draw down carbon dioxide from the atmosphere doesn't exist today, yet the 1.5°C pathway appears to rely on them. The decision text to the Paris Agreement has called on the IPCC to deliberate on this issue, with a special report requested by 2018.

During the countdown to the Paris COP21, the call for establishing the goal of net-zero emissions in the second half of the century reached a crescendo. Perhaps none were as loud as the calls made not by environmental

NGOs such as Greenpeace (which has published a plan to achieve such an outcome,[34] based on a huge reduction in energy demand on the back of efficiency and the rapid deployment of renewables) but by The B Team. This is a high profile group of business and civil society leaders, counting amongst its number Richard Branson (Virgin Group of Companies), Paul Polman (CEO of Unilever) and Arianna Huffington (Huffington Post). The team is looking at climate change, but also the challenge of doing business in the 21st century: shifting from Plan A, which requires business to focus on profit alone, to Plan B, which encompasses a more holistic set of objectives around financial performance, sustainability and business as a force for good to help solve challenging social and environmental goals. It is perhaps the next big step forward in what was originally termed 'sustainable development'.

Other than the call itself, what exactly is their big idea? Like many such goals, it is open to interpretation and raises questions as to how it can be achieved.

In a scenario that initially achieves stabilisation of carbon dioxide in the atmosphere, emissions could remain in the range 7–10 billion tonnes carbon dioxide per annum for some time without driving the atmospheric concentration higher because of the ocean's uptake of atmospheric carbon dioxide. Such a level of emissions is far below current levels (~33 billion tonnes per annum from the energy system alone), but it isn't zero. It could be classified as net-zero though, given that the atmospheric concentration isn't rising.

However, such an outcome, while stabilising the atmospheric concentration for a period, may not maintain stabilisation over the longer term and may not be sufficient to prevent dangerous interference with the climate system. In that case an even lower level of emissions may be required, such that atmospheric concentrations do begin to fall and stabilise at a lower concentration, which ultimately translates to a lower surface temperature, perhaps even as low as 1.5°C.

Scientists from the MITJP highlighted these issues in a July 2013 paper.[35] The MIT team deliberately avoided the use of net-negative technologies that draw carbon dioxide out of the atmosphere, partly due to concerns about their scalability but also preferring to test the impact of natural sinks on the outcome. Of these, the ocean is the major short-term natural sink because of the imbalance between levels of carbon dioxide in the ocean and the atmosphere.

Carbon dioxide dissolves in seawater. The gas will distribute itself between the ocean and the atmosphere, achieving a predictable equilibrium split for a given amount of carbon dioxide. But this process takes time, such that if additional carbon dioxide is added to the atmosphere, there will be a lag before a new equilibrium split results. With carbon dioxide being constantly added to the atmosphere, there is a permanent lag in ocean uptake of the gas. This means that if anthropogenic emissions cease, the ocean will continue to take up carbon dioxide and there will be an immediate fall in the level of carbon dioxide in the atmosphere. The decline in atmospheric concentration is relatively rapid in the early years after cessation of emissions as carbon dioxide levels in the mixed layer of the ocean[36] are more strongly out of balance with levels in the atmosphere. As the ocean's mixed layer comes into balance with the atmosphere, ocean uptake of carbon dioxide slows. Absorption by the deep ocean is a much slower process and will continue for hundreds to thousands of years.

MIT analysed four pathways that result in net-zero anthropogenic emissions. Of these, the pathway based on a more extended drop to net-zero emissions by 2060, with the decline commencing in 2010, is of particular interest. In this case anthropogenic emissions start to decline from 2010 when energy-related carbon dioxide emissions are at 30 Gt per annum (in reality they are at ~33 Gt by 2015). This scenario sees temperatures rise above 2°C by mid-century, but then decline as the ocean takes up significant quantities of carbon dioxide from the atmosphere with nothing being added from anthropogenic sources. After some 20-30 years, as the ocean's upper layer comes into balance with the atmosphere, uptake of carbon dioxide slows. There is of course a disturbing flip side to this story — continued rapid uptake of carbon dioxide by the ocean gives rise to increasing levels of ocean acidification already responsible for the climate-related death of coral reefs around the world.[37]

Back in 2010 when this theoretical pathway was deemed to commence, the cumulative emissions from 1750 (to 2010) stood at some 532 billion tonnes carbon. The area under their curve from 2010 to 2060 (energy, cement and land use) represents an additional 250 billion tonnes of carbon emissions, giving a total of some 780 billion tonnes. The modelled pathway results in an end-of-century temperature rise of approximately 1.5°C, which is also consistent with the relationship between carbon emissions and temperature of about 2°C per trillion tonnes.

The natural sink is therefore very important, potentially offering some 0.5°C of temperature reduction following an overshoot. This is possibly the only way in which 1.5°C can be met, although significant anthropogenic sinks may also be developed (including reforestation) to offer even greater drawdown.

The job of rapidly reducing global emissions hasn't yet started. Arguably, there are at least 40 + years to think about where the endpoint should be. In its initial formulation of a long-term carbon budget, the United Kingdom looked ahead to 2050, but that was from a 2008 starting point. With a new starting point of 2020 or thereabouts, a 2060 or even 2070 endpoint may well be considered. This could be another important output from the IPCC in their 2018 report.

A more stringent definition of net-zero emissions and the one that the Paris Agreement refers to is its application exclusively to anthropogenic emissions, irrespective of stabilisation or otherwise in atmospheric concentration. In this case, any remaining emissions from anthropogenic sources (and there will be some) would have to be offset with sequestration of carbon dioxide, either via CCS or a permanent forestry solution. In the CCS case, the carbon dioxide would need to come from a biosource, such as the combustion of biomass in a power station or through air capture.

The final step that goes beyond net zero is to have an anthropogenic net-negative emissions situation, which is drawing down on the level of carbon dioxide in the atmosphere through some anthropogenic process beyond that which natural sinks are able to achieve. The aforementioned BECCS and DACCS could enable such an outcome.

Finally, consideration needs to be given to those greenhouse gases besides carbon dioxide. The approach to take will depend on their longevity in the atmosphere. Methane, while a potent greenhouse gas, is relatively short lived (a decade or so) in the atmosphere so emissions can remain as long as the atmospheric concentration is stable or even falling. By contrast, nitrous oxide is both a potent greenhouse gas and long lived in the atmosphere. Even in a net-zero energy emissions system, nitrous oxide from agriculture will doubtless remain a problem.

The Paris Agreement has set the world on a new course in terms of a response to climate change and emissions mitigation. The ambition is huge, but the policy prescription is only just beginning to emerge. The solution lies primarily with individual nations, with the Agreement offering structure, transparency and a potential framework for cooperative action through the use of a market-based approach. This latter attribute of the Paris Agreement may well turn out to be its most important.

6

A GLOBALLY RELEVANT POLICY APPROACH

Humankind is effectively re-engineering the climate on a planetary scale over a relatively short space of time. The carbon dioxide emissions from the global energy system are some 35 billion tonnes per annum — the daily equivalent of a pyramid of carbon dioxide balloons, each containing one tonne, covering the island of Manhattan, with the tip of the pyramid at a height of about four kilometres.

Such is the scale of energy use in combination with unmet demand in many developing countries that any local attempt to reduce emissions, even if driven by a carbon price, may have little to no impact on the global picture. For both geographical and temporal reasons, the eventual accumulation of carbon dioxide in the atmosphere may remain unaltered.

While a carbon price is regarded as the most efficient means of driving change, it is also so efficient that it can lead to its own downfall. Local implementation of a carbon price skews local economics, which is manageable in the short to medium term as other locations implement similar pricing. But over the long term should others not take similar action, the economy will efficiently regroup around the local distortion. All other factors being equal, activities that are penalised through the action of the carbon price will progressively shift to areas where the penalty doesn't exist. This shift may explain why partial implementation of carbon pricing around the world has yet to have a visible impact on global emissions: inconsistent local implementation leads to a rearrangement of global activities, while global emissions continue without interruption, driven by increasing demands on energy supply.

A GLOBAL CARBON PRICE

While it is unrealistic to expect a carbon price to emerge globally without a hitch, over time that price must embed itself within the global economy so as to function effectively. Arguably, this embedding should be the single objective of a global approach to managing carbon dioxide emissions. The Kyoto Protocol didn't contain such a direct objective, but its approach involved price discovery through the trading of emission allowances, which encouraged the emergence of a global price. It even created a global currency, the Assigned Amount Unit (AAU), as well as the related Emission Reduction Units (ERU) of Joint Implementation (JI) and the Certified Emission Reductions (CER) of the Clean Development Mechanism (CDM).

Taken together, these forced a standardised approach to emission reductions and introduced a single carbon price into the global economy — or at least they were meant to. The AAU is similar to an allowance under a cap-and-trade system and is issued to participating governments with absolute targets under the Protocol. In theory, building domestic approaches on the back of the national assignment of AAUs meant that such systems could easily link up, with their domestic units exchanged for AAUs and vice versa. The commonality of the AAU also meant that nations could be quite inventive in implementing national action.

But this is the system that has now been tossed out, in spite of its carbon pricing design and structure. It lacked a mechanism to progressively expand absolute targets and AAU allocation to developing countries — so rather than trying to implement that expansion, political actors defeated the process. Nevertheless, as the Kyoto Protocol departs the scene, it leaves us with the legacy of carbon pricing, carbon markets, emissions trading and a demonstration of their collective effectiveness in shifting funds, triggering project activity and reporting on emissions. Some have argued that China would never be heading towards a national emissions trading system had it not been for the broad use of the CDM in that country and the commercial interest it stirred.

I often thought that it would be much simpler for countries looking to implement carbon-pricing systems to just join the EU ETS; after all it has been there since 2005. A few EU neighbours did join (e.g. Norway). But governments don't tend to make simple practical decisions; rather, they prefer to design nationally specific approaches. Both Australia and the EU were signatories to the Kyoto Protocol in the period 2008–2012, so their national emissions were already governed by AAUs, which meant that any kind of sovereignty issue shouldn't exist. As such, attaching some or all of the Australian emissions reduction effort to the EU ETS would have been relatively easy. But that wasn't the path that was followed. Rather, Australia went through several attempts to design a domestic system, which they eventually did, with the end result being quite different from cap-and-trade.

About two years before the Direct Action emissions management system was implemented in Australia, the Australian government and the European Commission announced that their respective emission trading systems would join together, or 'link', to use the climate vernacular, over Phase III of the EU system (between 2013 and 2020). This was a bold move by both parties — perhaps they were hoping that such a move would encourage others with nascent trading systems to sit up and think about where they wanted to go. For Australia, the move was also designed to give local credibility to their new system and for the EU it could give new life to the beleaguered ETS. A full two-way link between the two cap-and-trade systems was planned to start no later than July 2018.

Although the EU-Australia link was never to be following the termination of the Australian system by the government of Tony Abbott, at the time the move raised a key issue: namely, the future design by the UNFCCC (or other entity) of any international framework. Both Australia and the EU stressed that theirs was a bilateral linkage, to the extent that they would not use the tools offered by the UNFCCC to record their cross-border allowance transactions. However, both were signatories to the Kyoto Protocol, so regardless of their claims of bilateral policy development, there would still have been

Kyoto AAU transactions at various times to ensure compliance in that system and to keep it whole.

Despite their apparent distancing from the Kyoto-based International Transaction Log (ITL), it is still the case that the overarching Kyoto Protocol framework provided the foundation for this proposal — perhaps even allowing it to happen. Thanks to the UNFCCC architecture, the two systems grew up successfully enough to make a linkage possible. They counted allowances the same way, tracked emission units the same way and had similar compliance requirements. Both the systems had common offset arrangements through CERs under the CDM, and the units created under the Australian Carbon Farming Initiative were also Kyoto-compliant. In other words, because of the Kyoto architecture, the makings of a linked system were within reach.

Linkage of carbon-pricing systems should be a priority for any new international emission-reduction framework. Such a framework would provide tools, rules and mechanisms that countries could use to develop their carbon-trading systems, thus facilitating and even encouraging linkage at a convenient time for those interested in doing so. Providing a framework for linking could eventually deliver the global market that is needed.

Carbon pricing nearly fell off the international agenda. Following the failure of COP15 in Copenhagen and the bitter taste of market-based approaches in the wake of the global financial crisis, movement towards implementing governmental carbon pricing systems went into decline in 2010 and 2011. In June 2012 the World Business Council for Sustainable Development (WBCSD) and the International Emissions Trading Association

(IETA) convened a crisis meeting. Amongst other things, the meeting gave rise to the Carbon Price Communiqué that was launched later that year by the Prince of Wales Corporate Leaders Group (CLG) within the United Kingdom. This provided a useful springboard, but considerably more had to be done to resurrect the carbon-pricing discussion.

Enter the OECD and World Bank. As the global political system began to prepare for COP21 in Paris, the concept of government-led carbon pricing was struggling. By May of 2014 when a preparatory meeting was held in Abu Dhabi for the September New York Climate Summit, carbon pricing didn't even make the final agenda. There was a veiled reference to it under the heading 'Economic Instruments', but such a phrase could mean any number of things. Nevertheless, under the auspices of the OECD, World Bank and with WBCSD, IETA and now CLG offering strong business support, the tone changed. The Bank launched its Carbon Pricing Statement, and the UN Global Compact stepped in with a set of Carbon Pricing Principles; together they gained widespread support.

These efforts didn't fully rescue the situation, but they started a new government-led conversation about carbon pricing that ultimately led to the creation of the World Bank's Carbon Pricing Leadership Coalition (CPLC) in April 2016. The CPLC is a coalition of governments, NGOs and businesses working together to promote carbon pricing and encourage the implementation of systems and approaches at national level.

It might seem extraordinary that such actions were needed after 25 years of UNFCCC meetings, the Kyoto Protocol and implementation of various carbon pricing systems, supported by the scientific facts — but such was the state of the international process on climate change and the continued resistance to wholeheartedly embracing carbon pricing.

Although the agreed language of the Paris Agreement is NDCs, these may do little to stem emissions if a cost on carbon emissions isn't at the core of the national efforts underpinning the contributions. While a number of countries have developed or are implementing such policies, and the efforts of groups such as the CPLC may accelerate this process, the current carbon-emissions cost, as seen by an emitter in economies where implementation is underway, is unlikely to make a substantial difference. In 2016 the global average in countries using carbon pricing was probably less than US$10. While these efforts to establish carbon pricing are essential in terms of building the necessary national institutional capacity for a carbon price in the economy, there is little evidence that governments, business and consumers are prepared to accept the level of cost that will deliver a tangible change in energy investment. Both governments and the private sector are concerned about early adoption of carbon pricing due to competitiveness concerns. Put simply, a carbon price introduces a cost in an economy that may not be seen in other economies.

EXPANSION VIA LINKING

Arguably, the only way to meet these concerns is to create widespread adoption of carbon-pricing policies and some form of linkage between national systems to create price harmonisation. Harmonisation removes the concern of competitive disadvantage, both at a company level and at a national level.

Although the EU ETS has a single design, the initial period of operation saw EU Member States establishing their own allocation, so the system was in some senses a common approach built on linked independent systems. At the beginning of Phase II in 2008 the system expanded its reach with Norway, Iceland and Liechtenstein joining, which is the route Australia could potentially have taken rather than attempting to design its own system from the ground up and then negotiating the terms and conditions for linking it to the EU ETS later on. However, we are clearly in a phase of global climate policy development where independently designed and implemented national approaches are the agreed way forward, which means that the only route open to create a homogeneous global market is to link these various approaches together one by one. The glory of the first real international linkage between emission trading systems went to Canada and the United States, when the Province of Québec and the State of California joined their systems in 2014. But a much more comprehensive approach will eventually be needed.

The World Bank carbon markets team has also been hard at work developing a novel approach to linking disparate systems such as the EU ETS, the ETS that is now operating in South Korea and the various emissions trading systems operating in North America. This could either be impossibly difficult to implement or may even revolutionise the global carbon market — at this stage it is hard to assess which end of the spectrum might be arrived at. Nevertheless, it is an idea with real merit and worth exploring and even piloting.

The Bank takes the view that despite the best of intentions, market-based emissions-management systems (such as cap-and-trade or baseline-and-credit) will only rarely come close enough in design and underlying ambition to seamlessly link together. Further, as systems with existing bilaterally agreed linkages try to foster links with others (and therefore connect the system that they are already linked with to another one by default), progress will grind to a standstill. Therefore, something else is needed. The Bank's idea is to introduce a ratings system for individual market-based instruments, similar to the sovereign ratings system and handled by the likes of Standard and Poor's.

For example, a tight cap-and-trade system with limited offset use and high ambition (i.e. a sharply declining cap) might have its allowances rated at 0.9 (like a national AAA or AA + rating), compared with a baseline-and-credit system with credits rated at 0.3. In this particular example, the big difference is created because the cap-and-trade system deals with a finite level of carbon emissions, but the baseline-and-credit only limits emissions per tonne of production. Trade between the two would then be possible, but three external credits would be needed for compliance instead of one internal allowance in the cap-and-trade system. Many different systems could then link without the need for perfect alignment of their initial designs. Ratings applied in this way could potentially solve the problem that the EU Commission has had with it's on/off approach to CERs from the CDM.

In the World Bank model the ratings would be handled by a private agency, and the decision to use them would be a sovereign one, both by a country hosting a market-based system, which wishes to import other instruments for compliance and by a country that wished to export its carbon instruments for external use. Challenges remain with this approach, including acceptability of rating systems by sovereign nations and the degree to which such a system would work with fast-moving commodity markets. The idea is designed to commence outside the UNFCCC process and the World Bank continues to develop it.

But is a system of linked systems suitable for all countries, given the very wide spread of absolute levels of economic development? The Kyoto Protocol sought to address this issue through the principle of Common but Differentiated Responsibilities (CBDR), which, at the time, was interpreted to mean that developed countries should take the lead in reducing emissions and then assist developing countries adopt a low-emissions pathway on which to develop.

The countries which were classified as developed at that time — the United States, Canada, Norway, European Union, Switzerland, Japan, New Zealand and many of the former Soviet States — each adopted a specific emissions goal for the initial compliance period of 2008 to 2012, relative to a 1990 baseline. The goal was administered through the issuance by the UNFCCC of tradable AAUs to national governments. For example, if national emissions in 1990 for a given developed country were 1 billion tonnes and a reduction goal of 10% was agreed, then 4.5 billion units would be issued to that country for the five-year period. The compliance obligation on a national government was to surrender AAUs against actual emissions for that five-year period starting in 2008. Should national emissions of carbon dioxide and other greenhouse gases exceed the number of allowances issued to the country, then that government would have to purchase additional units from others; or if a surplus resulted, a country could sell its allowances to others or hold onto them for future compliance periods.

By doing this, the Kyoto Protocol introduced the rudiments of carbon pricing through a simple international cap-and-trade system. It was envisaged that AAUs would be scarce and therefore have value, which would manifest as a cost of emissions when the units were traded as nations looked towards their compliance obligations. Further, the system encouraged national governments to mirror their compliance risk within the economy by repeating the process and implementing cap-and-trade systems over their emitters, such as factories, power stations and even vehicles. By contrast, if a government chose to introduce different policies at a national level, for example by encouraging energy efficiency or supporting renewable energy as a means to reduce emissions, they would have a mismatch between internal and external approaches and therefore carbon price exposure.

One large-scale example of mirror implementation of the national obligation came from Europe where the EU ETS was put in place to cover all large-point source emissions across the EU, primarily, power stations and industrial facilities. This policy framework effectively took the provisions of the Kyoto Protocol and transferred them to the private sector to implement for nearly half of Europe's emissions. Although the EU ETS started in 2005, three years prior to the first Kyoto period of 2008–2012, the initial years were seen as a trial stage and were financially isolated from the formal start that took place in 2008 to match the Kyoto obligation.

GLOBALLY TRADED OFFSETS

The Protocol had a number of other provisions, but one in particular proved enormously popular with the private sector and was designed specifically to help developing countries. This was the CDM. Unfortunately, the CDM also contributed to the unravelling of the whole system over the years to come due to the scale of its use compared with the smaller-than-anticipated scale that the rest of the Kyoto Protocol operated on. The CDM appeared as just a few lines of text in the original Protocol document, but the detailed rules under which it eventually operated expanded to fill many volumes. The CDM transferred money for clean energy investment between developed and developing countries on a substantial scale, but also led to accusations of abuse on the part of certain entities.

The CDM is an offset mechanism. Offsets are variously described as carbon credits, emission reduction units and project mechanisms, but in all cases are measured carbon reductions from a specific activity that are applied to negate the emissions of another activity. For example, a wind farm is proposed for installation in a particular location, instead of a less expensive power plant that would have emitted carbon dioxide. The project developer needs additional investment to cover the added cost of the wind farm, so they turn to the offset mechanism. Certificates for the annual difference in emissions between the wind farm that is built and the emitting plant that wasn't built are awarded to the project developer by a government-approved agency that is assigned the task of assessing the project. These certificates can then be sold and used by another party to negate the same quantity of emissions in other circumstances, for example negating the need to buy a quantity of allowances in a cap-and-trade system. The proceeds from the sale boost the revenue of the wind farm, which in turn justifies the additional up-front investment.

The CDM has become the best-known and most widely used offset system. But it is also a financial mechanism that is designed to channel money from developed to developing countries for clean energy development purposes. A country such as Japan with a cap under the Kyoto Protocol can expand its national allocation either by buying AAUs from another country with a cap (therefore reducing that cap by the same amount), or it can effectively expand the cap of the entire Kyoto system by investing in a clean energy or emission-reduction project in a developing country through the CDM and import the reductions made as units known as CERs. CERs trade globally; this gives rise to a secondary carbon price that is linked to the prevailing price in developed countries that the CERs are destined for.

Offset mechanisms also appear in the public domain, such as those that have been used by airlines to offset the emissions from a particular flight. This is normally an optional cost for the traveller, which you may see as a check box when you come to pay for the flight online. This also introduces a form of carbon pricing into the mainstream public vocabulary. The airline uses these proceeds to buy enough certificates from clean energy or reforestation projects to offset the emissions associated with the use of the jet fuel to fly the plane.

A feature of these systems is that the accounting normally handles the entities within the cap and the project outside the cap, but no attempt is made to account for the total greenhouse gas impact on the atmosphere or against a global goal to reduce overall greenhouse gas emissions. There is an implicit assumption that the sum of the various parts adds up such that the overall outcome is better than not having conducted the exercise at all. The assumption is necessary because only a small percentage of the global economy sits under a cap, so there is no mechanism available to account for the total impact. The lack of more comprehensive accounting is one reason why some parties challenged the appropriateness of the Kyoto Protocol itself.

In order to address the accounting issues, offsets have to pass a test known as 'additionality'. For the offset to be valid it should represent a real reduction that would not have taken place in the absence of the offset project. The question is: would the project have happened anyway? Is it in addition to what would have been the normal course of events? Furthermore, does the offset represent a real reduction?

Like almost all emission-reduction activities seen in the world today, offsets still tend to be judged on the basis of local emission changes rather than their contribution to the eventual atmospheric stock of carbon dioxide. For example, a small-scale project that sees solar PV installed on village roofs in a town in Uganda is a worthwhile venture, but determining its contribution to limiting the eventual concentration of carbon dioxide in the atmosphere doesn't get much attention. The additionality assessment focuses on calculated avoided emissions at the time of implementation, which may have little link to the future cumulative emissions as a result of Ugandan development. Nevertheless, that is the test under the CDM.

All current emission reduction systems tend to gloss over the cumulative emissions issue, although in the offset world those that encourage reforestation are perhaps closest to delivering a real reduction, assuming of course that the new forests become a permanent feature of the landscape and can be shown to represent a real increase in global forest coverage. But the rest just assume that reducing current local emissions or avoiding future emissions equates to an eventual shift in cumulative emissions. This may not be the case because of the previously discussed coal miner who will continue to extract the accessible resource until it is depleted, or the solar PV installation which in the interim has encouraged the villagers to buy refrigerators and quite possibly advanced the date of the first village grid connection. That grid may even be powered by the abundant coal resources in the region.

Solving this transition problem relies on the underlying implication of CBDR; that those countries with high cumulative historical emissions (the developed countries) should offer financial help to others (the developing countries) so that they can develop a clean energy pathway rather than relying on fossil fuels, particularly on coal.

Enter the CDM. The idea here was to leapfrog over the coal era and move directly to natural gas or renewable energy by providing a subsidy for more advanced energy infrastructure. This subsidy would arise through the sale of CERs within trading systems such as the EU ETS or directly to developed-country national governments needing to sharply reduce their emissions but unable to do so domestically to the extent required. But the CDM hasn't quite worked out that way. With limited buyers of CERs and therefore limited provision of the necessary subsidy, the focus shifted to smaller-scale projects, such as the rural electricity project in Uganda. While projects like these are good for the communities involved, they don't represent the necessary investment in large-scale industrial infrastructure that a country needs to develop. Rooftop solar PV won't build roads, bridges and hospitals or run steel mills and cement plants. So the economy turns to coal with the smaller projects failing to deliver a real reduction in emissions. In fact, they may hasten development and result in big infrastructure arriving sooner rather than later — which would constitute an emissions increase!

With the proliferation of the smaller CDM projects (along with the number of companies developing them), the limited CER outlets were flooded with the reduction units, adding to the downward price pressure that the CER was already experiencing. Some argue that the influx of CERs was a significant contributor to the collapse of the EU ETS carbon price. That influx also meant a collapse in the CER price to levels well below that experienced by the ETS because the EU was eventually forced to close the door and therefore shut off demand almost completely. In 2010 the CER spot price traded between €10 and €15, but by early 2013 it was trading at just a few cents.

In hindsight, the decision to focus on smaller projects may have taken the CDM in the wrong direction. Rather, a focus on large-scale projects, say greater than 2−3 million tonnes per annum of carbon dioxide, might well have restricted the flow of CERs to the EU and therefore raised their price. In turn these CERs could have financed large-scale industrial and power generation projects with an emphasis on least developed economies where coal lock-in and subsequent emissions accumulation had yet to occur or was just starting.

There is no doubt that offset mechanisms operating around the world today have channelled significant funds between countries in the case of the CDM and between sectors within countries in the case of domestic systems such as the offsets allowed under the California cap-and-trade system. They remain an important feature of emissions-trading system design in that they offer great flexibility to those operating within such a system. But the question remains as to their real ability to reduce emissions and shift development pathways. While local reductions have been measured and claimed, these do not necessarily translate into reductions in global cumulative emissions. The link is more tenuous than between a system such as the EU ETS and similar long-term reductions.

Only the EU (and to be fair, New Zealand) persisted with full implementation, passing at least some of the carbon-price exposure it had adopted under the Kyoto Protocol into the economy by implementing a domestic emissions trading system and then linking that system back with the Protocol via the CDM. In combination with significant Japanese buying of CERs this allowed the CDM to channel some $50–$100 billion in project investment to developing countries.

The CDM succeeded to an extent, although detractors continued to criticise it. Claims of rent seeking[38] abounded, with accusations of chemical companies in developing countries profiting from the mechanism through their programme of HFC-23 destruction rather than HFC-23 venting to atmosphere. HFC-23 is a greenhouse gas with a very high carbon dioxide equivalence factor and is an unwanted by-product in the manufacture of HCFC-22, a refrigerant. The destruction of HFC-23 in HCFC-22 plants in developing countries can be registered as a CDM project and leads to the issuance of a large number of CERs. As it is very cheap to install a destruction facility, HFC-23 destruction CDM projects have resulted in significant profits for HCFC-22 plants and, some detractors have argued, a perverse incentive to increase the production of HCFC-22 (although this was at a time of a huge expansion of refrigeration capacity in China). Nevertheless, the mechanism worked. HFC-23 was destroyed, and changes took place within the CDM and the CER recipient countries to curtail profiteering. While it is probably true that some rent seeking went on, the CDM nevertheless inspired a range of positive activities, from landfill gas capture to renewable energy installation. But in doing so and with just a handful of buyers, the system soon flooded the limited market with CERs, and the price plunged.

CARBON PRICING AND THE PARIS AGREEMENT

The Kyoto Protocol has all the necessary elements of a system that could have delivered a global carbon price, but its reach was never broad enough. The CDM is an opportunity-based mechanism, meaning that it only applies when the financial opportunity exists, and is therefore ineffectual when it comes to the broader energy system. It should not be left to form the basis of a fully fledged carbon price for the majority of developing countries, but rather should be left to focus on real development pathway shifts in poorer economies.

A solution for the limited reach of Kyoto Protocol could have been formulated had negotiators not stuck so rigidly to the developed/developing country split as it existed in 1992. A means to promote emerging economies into the group that had absolute targets was needed, but no such proposal was ever introduced. The very notion ran headlong into the founding principles of the UNFCCC that implied permanent differentiation between the developed and developing countries. So the conversation moved on.

Surprisingly, a concept along such lines actually appeared at COP20 in Lima, authored by the Brazilian delegation, but it was introduced as a proposal for the new global deal to be agreed in Paris in 2015 rather than as a fix for the Kyoto Protocol. Brazil has a long history of creative intervention in the process, being the country that initially proposed the CDM. It appears to me that Brazil returned to its creative best with a November 2014 submission to the UNFCCC entitled 'Views of Brazil on the Elements of the New Agreement under the Convention Applicable to All Parties'. The proposal still recognised the difference between developed and developing countries but also recognised the need for universal acceptance and eventual implementation of absolute targets, including for developing countries, as the route to atmospheric stabilisation of carbon dioxide.

This means that every country would migrate towards absolute targets as its capabilities allowed, but that developing countries would at least start with an emissions goal,

albeit intensity based, per-capita based or based on a business-as-usual deviation. Least developed economies are
encouraged, but not required, to present a contribution.

The Brazilian proposal also included extensive references to markets, cap-and-trade, a reformed CDM and so
on, but it was probably a decade late in being proposed,
and it was not accepted in Paris. Rather, the negotiators
at COP21 took a more circuitous route towards carbon
pricing and the emergence of a global carbon market:
they gifted the world with Article 6 of the Paris
Agreement.

Article 6 was something of a miracle, with negotiators
reporting that there was no carbon pricing or carbon-
market language in the draft text as late as the day prior
to the final adoption of the Paris Agreement. There was
certainly a negotiating group working towards such a
goal, but its efforts were not always recognised by the
broader plenary group assembled at COP21.
Nevertheless, a last minute push saw the text inserted,
not just as a few lines, as many expected, but as a comprehensive Article that would feature heavily within the
overall Agreement.

Specifically, it is Paragraph 2 of Article 6 that pertains
to potential linkage between NDCs, such as might occur
between cap-and-trade systems operating in different
countries. Take the case of the linkage between the
California and Québec systems. Presumably, when
Canada presents the result of its NDC efforts to the
UNFCCC at various stocktakes, any net transfer that has
occurred between these two systems will need to be
accounted for, with this paragraph leading to a clear set
of modalities for the necessary accounting of the transfer.

The longer-term hope is that this paragraph provides additional impetus to such activities, catalysing both the use of trading systems and the creation of links between them. This is an important step towards the formation of a globally traded carbon market.

Paragraph 4 of Article 6 is also a formative proposition, potentially containing within it the means to drive new investment and markets. It creates a mechanism to contribute to the mitigation of greenhouse gas emissions and the support of sustainable development. One early reading of this paragraph is that the mechanism becomes the CDM of the Paris Agreement, i.e. CDM 2.0, but the text does not mention project activity or identify developing countries as the beneficiaries of the activities undertaken. This is in contrast to Article 12 of the Kyoto Protocol, which clearly identified such a role for the CDM. But more importantly, such a simplified structure isn't possible under the Paris Agreement because of the role of NDCs and the essential requirement to avoid double counting.

The Article 6 outcome only happened thanks to many months of advocacy and legwork. IETA were the ones who picked up on this issue, releasing a straw-man proposal for the Paris Agreement during the 2014 New York Climate Summit.

IETA imagined a relatively short Paris Agreement that devoted just a few paragraphs to each key subject. In reality this was very close to the mark, with the longest section of the Paris Agreement covering the subject of adaptation to climate change. It was also clear that a focussed proposal on carbon pricing or international emissions trading would not make the cut, so a more tangential approach would be needed to include these key concepts in the final text. The Agreement could offer the platform on which various national carbon-pricing policies could interact through linkage, bringing a degree of homogeneity and price alignment between otherwise disparate and independently designed systems. The case for linkage was initially put forward through collaboration between IETA and the Harvard Kennedy School in Massachusetts. A number of papers coming from the school underpinned the IETA straw-man proposal for the Paris Agreement. The proposal didn't mention carbon pricing or emissions trading; instead it offered a provision for transfer of obligation between respective NDCs, in combination with rigorous accounting to support the transfer.

> ... *may transfer portions of its defined national contribution to one or more other Parties* ...

In addition, the proposal suggested a broader mechanism for project activity and REDD + .[39] The IETA team worked hard during 2015 to build the case for such inclusions in the Paris Agreement. A number of governments, business groups and environmental NGOs reached a similar conclusion: Paris needed to underpin carbon-market development. After all, fossil fuel use and carbon emissions are so integrated into the global economy that only the power of the global market could usher in the necessary transition.

Article 6 of the Paris Agreement now provides the opportunity for NDC transfer between parties and a mechanism to contribute to the mitigation of greenhouse gas emissions and supporting sustainable development. The transfer provision is described in Article 6 as follows:

> ... *approaches that involve the use of internationally transferred mitigation outcomes towards nationally determined contributions ...*

While not exactly the same as the original IETA idea, it does the same job. Of course, like every other part of the Paris Agreement, this is just the beginning of the task ahead. The CDM within the Kyoto Protocol was similarly defined back in 1997, but it was not until COP7 in Marrakech in 2001 that a fully operational system came into being. Even then, the CDM required further revisions over the ensuing years.

The Paris Agreement is built on the concept of Nationally Determined Contributions (NDCs). These are set at a national level and offer a direction of travel for a given economy in terms of its energy mix and/or greenhouse gas emissions. Although the first set of NDCs offered in the run-up to COP21 were varied in nature and in some cases only covered specific activities within the economy, over time, they will likely converge in style and, for the Paris Agreement to deliver, must expand to cover all anthropogenic greenhouse gas sources.

A NEW GLOBAL ACCOUNTING REGIMEN

The NDCs also lead us down another path — that of quantification. The first assessment of NDCs conducted by the UNFCCC in October 2015 and then refreshed in May 2016 required the quantification of all NDCs in terms of annual emissions and cumulative emissions through to 2030. This quantification was necessary to establish an equivalent level of warming of the climate system, which is driven largely by the cumulative emissions of carbon dioxide over time. Without such an assessment, the UN cannot advise the parties on progress towards the aim of the Paris Agreement.

The UNFCCC didn't have a full emissions inventory on which to base this calculation, so they established one from the best data available. But Article 13 of the Paris Agreement introduces a transparency framework and calls on parties to regularly provide:

- *A national inventory report of anthropogenic emissions by sources and removals by sinks of greenhouse gases, prepared using good practice methodologies accepted by the Intergovernmental Panel on Climate Change and agreed upon by the Conference of the Parties serving as the meeting of the parties to the Paris Agreement;*

- *Information necessary to track progress made in implementing and achieving its nationally determined contribution under Article 4.*

The foundation for transparency is measurement and reporting, which further implies that emissions quantification is a foundation element of the Paris Agreement. Although NDCs are nationally determined and always voluntary, the Agreement effectively establishes a cap, albeit notional in many cases, on emissions in every country. The caps are also effectively declining over time, even for countries with emissions still rising as development drives industrialisation.

Article 6 introduces the prospect of carbon unit trading through its internationally transferred mitigation outcomes (ITMO) and emissions mitigation mechanism. Paragraphs 6.2 and 6.5 make clear that double counting is not allowed under the Agreement:

> ... *internationally transferred mitigation outcomes towards nationally determined contributions ... shall apply robust accounting to ensure, inter alia, the avoidance of double counting,*

> *Emission reductions resulting from the mechanism referred to in paragraph 4 of this Article shall not be used to demonstrate achievement of the host Party's nationally determined contribution if used by another Party to demonstrate achievement of its nationally determined contribution.*

These provisions, in combination with the progressive shift towards quantification of all emission sinks and sources, means that full national accounting for offset crediting must take place for both the recipient and the source of the units. For the recipient, there will be no change in their procedures in that the introduction and counting of outside units is already built in to the inventory processes underpinning the trading systems. But the source country will be required to make an equivalent reduction (also referred to as a corresponding adjustment) from their stated NDC, therefore tightening their contribution. This was not the required practice in the CDM, but under the Paris Agreement everything changes.

The CDM utilised the more subjective additionality test, but under the Paris Agreement there should be no further need for this as both sides of the transaction will be required to account for the transfer in their national emissions inventory.

For example, consider the hypothetical case of a nature-based transfer from Kenya to Canada, utilising Article 6 as a means to acquire the necessary funding.

- Canada is implementing a reduction of 30% by 2030, with 2005 as the baseline. Therefore, over the period 2020−2030 it has an effective cumulative emissions cap of 5650 Mt.

- The Kenyan economy is industrialising and therefore expects to see its emissions rise significantly by 2030, but it pledges to keep its emissions 30% below this level. Although this is a notional trajectory, it can nevertheless be formulated as a 1000 Mt cumulative emissions limit over the period 2020−2030.

- As part of a suite of measures, Kenya intends to increase forest coverage. It sells 50 Mt of carbon dioxide reductions as a nature-based transfer to Canada as a means of raising finance for its reforestation efforts.

- The transfer impacts both NDCs. Canada raises its cumulative cap to 5700 Mt for the period, but Kenya must now adjust its domestic ambition, targeting instead a 37% reduction below its expected emissions growth. This ensures there is no double counting of the transferred amount and maintains the full integrity of the overall NDC approach such that the implied global cumulative emissions goal of the NDCs is maintained.

Kenya will need to find further reductions in its economy as a result. One implication of this is that the price of carbon units may rise due to the additional demand that an overall emissions cap, even a notional one, places on the global economy.

Article 6 of the Paris Agreement offers great potential for carbon-market development and emissions trading, therefore driving a lowest-cost mitigation outcome and directing funding and financing to low-emission technologies. But over time, it should also introduce an accounting rigour that until now has featured only in some quarters. This rigour may well change the supply-demand balance, leading to a more robust and enduring carbon market.

The Paris Agreement offers a global framework within which countries develop and operate their own policies and goals. The previous three years saw governments, business and civil society organisations work collaboratively to build the foundations upon which an agreement regarding a global carbon market could also be established.

In the end, far more was achieved than any of us seriously imagined. The outcome is a win for the global economy in terms of the long-term cost management of rectifying the climate problem. And it is also a win for the environment.

The ambition embedded within the Paris Agreement is going to require change on a very large scale and at a very rapid pace and certainly much faster than could be envisaged through a project-by-project approach, as was the case with the CDM. While the CDM was successful in what it did, the scale was relatively insignificant given the size of the global energy system. This also argues against a narrow use of the mechanism described in Article 6.4. Rather, it should have very wide scope and operate on many fronts, potentially functioning as a universal carbon-trading unit. Other definitions or uses may also be considered.

Whether Article 6 is a sufficient enough catalyst to build an entire global carbon market remains to be seen. Like the rest of the Paris Agreement, its success will be determined by the rigour shown by governments as they approach the task of implementation. But some early signs have emerged, including the aforementioned meeting between President Obama and Canadian prime minister Justin Trudeau. Most important was the recognition that co-operative action is required to implement Article 6. The joint statement released during the meeting made a very specific reference to the required implementation with the commitment that both countries would work together to support robust implementation of the carbon markets-related provisions of the Paris Agreement. The statement noted that the work would explore options for ensuring the environmental integrity of transferred units, in the context of NDC accounting and efforts to avoid double-counting.

The statement represents a big step forward for the United States and for the further development of carbon markets. The United States was amongst the very first countries to release its NDC, but noted within it that it did not intend to utilise international market mechanisms to implement its 2025 target.

This did not come as a surprise given that it was early days for the resurgent political interest in the importance of government implementation of carbon pricing and, by association, the supporting role that international carbon markets can play in helping optimise its use. But a great deal had happened in the intervening year (the United States released its NDC on 25 March 2015), topped off with Article 6 in the Paris Agreement.

A further year on and there is uncertainty again at the Federal level in the United States with a new administration taking office, but it is the developments at state and provincial level that have raised the profile of cross-border carbon unit trade. At the April 2015 Québec Summit on Climate Change, Ontario announced its intention to set up a cap-and-trade system and join the Québec-California carbon market. The following September, Québec and Ontario signed a cooperation agreement aimed at facilitating Ontario's potential participation. To add to this, during COP21 Manitoba announced that it would implement, for its large emitters, a cap-and-trade system compatible with the Québec-California carbon market. Then in Paris, Québec and Ontario committed to collaboration with Manitoba in the development of its system by signing a memorandum of understanding to that effect. The Ontario cap-and-trade system started in January 2017 and is expected to link with the California-Québec system in 2018.

Other US states and Canadian provinces may join, with Mexico also expressing interest, leading in turn to a significant North American grouping of carbon markets, perhaps even one starting to match the scale and breadth of the 30 member EU ETS. Grouping of regional carbon markets into so-called carbon clubs is seen by many observers as the quickest and most effective route to an eventual global market.

With parts of the United States and Canada forming regional carbon markets, the respective national governments are effectively bound to build the existence of these markets into their NDC thinking. There may be a significant flow of units across national borders, which will make it necessary to account for them through Article 6 and the various transparency provisions of the Paris Agreement. But most importantly, there is economic benefit in linkage and transfers; a larger more diverse market will almost certainly see a lower cost of carbon across the participating jurisdictions than would otherwise have been the case.

7

ENGINEERING THE TRANSITION

While there were many examples of high-level international diplomacy in preparation for COP21 in Paris, it was a visit by President Obama to India and his discussions on climate change with the Indian prime minister Narendra Modi that threw the spotlight on the global development pathway and its energy needs.

The story that grabbed my attention wasn't about the president and the prime minister, but about Santosh Chowdhury, a gentleman who lives in the village of Rameshwarpur in Orissa, an eastern state of India. According to the BBC report, Mr Chowdhury had just bought a refrigerator, which may seem unremarkable — but it was the first fridge in his village. Now the thing about a refrigerator is that, unlike many other domestic appliances, it requires a reliable electricity supply 24 hours a day, 365 days a year. This means that Mr Chowdhury needs a grid connection capable of sending electrons his way at any time throughout the day.

Because the fridge needs electricity on a continuous basis, this excludes local intermittent renewable energy as the primary provider unless a reliable backup or energy storage mechanism is available. In India, given cost considerations, the baseload electricity that guarantees reliable delivery will likely be generated with coal although it is clear that India is also looking towards nuclear energy. Solar energy will augment this and at certain times may provide for all Mr Chowdhury's needs. But unless the town spends considerably more money and installs a more complex micro-grid system with battery capacity, the dependency on coal will continue, at least in the medium term.

But the story doesn't end there. The lack of fridges in Rameshwarpur reflects the situation across the whole of India. In 2016 only about one in four of the country's homes contains a fridge. That compares to an average of 99% of households in developed countries. In 2004, 24% of households in China owned a fridge. Ten years later this had shot up to 88%, or some 400 million refrigerators. India has about 250 million households, which approximates to 60 million fridges today. By 2030 as population rises, as people per household likely decline and as fridge ownership approaches Chinese levels, India might have 400 million fridges.

So Mr Chowdhury's purchase and many others following will mean that India needs to produce more fridges — over 300 million more. This fridge is the start of a long industrial chain. It means India needs more refrigerator factories and chemical plants to make the refrigerant. The refrigerators will be made of steel and aluminium, which means mining or the import of ores, refining, smelting, casting, stamping and transport. All of these need coal, gas or oil. Coal in particular is needed for smelting iron ore as it acts as the reducing agent, producing carbon dioxide in the process. The intense heat required in the processes is most easily and economically provided by coal or gas, although given time electricity will doubtless make its way into these processes.

The refining and fabrication of raw materials required by developed and developing countries' secondary industry demands considerable use of fossil fuels for combustion-based processes such as smelting, refining and base chemical manufacture. Even if we imagined a world reliant on 3D printing, which uses exotic materials (graphene, certain polymers, etc.) as the raw material for manufacture, considerable chemical plant capacity and process heat would be required to manufacture the printer feedstock.

Then these materials must be shipped from mines to refineries to manufacturing plants to distribution depots, and then to wholesalers, to shops and finally to Mr Chowdhury's home. Although electricity is starting to appear in the transport sector for lighter vehicles, with the exception of railways, it isn't yet the energy provider for heavy transport. In India, rail transport is extensive, and while electrification is making good progress, there is still a large role for oil.

With a refrigerator, family life for the Chowdhurys will undoubtedly improve. It will be easier to manage food supply and meal preparation, so productivity in other areas may well rise. This could translate to more income, further purchases and perhaps the first opportunity for air travel, which will almost certainly be powered by a hydrocarbon fuel.

China has now reached such a point, with the aviation sector expanding exponentially. In a September 2016 update to shareholders, Boeing reported that Chinese airlines are expected to order some 7,000 planes at a cost of US$1 trillion through to 2035 along with the need to train some 100,000 pilots.

There is no doubt that India is industrialising rapidly, and Prime Minister Modi should be commended for his ambitious goal of 100 GW of solar capacity by 2020 and speeding up the nuclear programme — but this won't stop carbon dioxide emissions from rising sharply in the near term. It is more a question of how high they will rise and the more immediate actions that can be taken to limit the rise.

On the back of this rapid development, India is reported[40] to be opening a new coal mine every month. Its goal is to produce 1.5 billion metric tons of coal by 2020, exceeding US coal production and becoming the second largest coal producer in the world, after China. Depending on the quality and ash content, 1.5 billion tonnes of coal is equivalent to some 4 billion tonnes per annum of carbon dioxide, or about 12% of current global energy-related emissions.

RATES OF CHANGE

The Paris Agreement may be startling in its simplicity and huge on ambition, but the question is, does it guarantee success in tackling global emissions, particularly as countries such as India move into a rapid development cycle such as seen in China so far during this century?

In the back pages of the October 2013 edition of *The Economist*, two Indian states called for interested parties to pre-qualify for the construction of 8 GW of coal-fired power stations — just two examples of similar cases taking place across the world to bring much-needed electricity to rapidly emerging economies. These power stations, once built, can run for up to 50 years, delivering a total of 2.5 billion tonnes of carbon dioxide to the atmosphere or nearly 700 million tonnes of carbon.

In this particular case, the power plant will deliver about the same amount of electricity as all the installed (up to 2013) wind and solar capacity in India. That's about 22 GW, but with a load factor of around 0.3, gives something similar to 8 GW of coal.

While India has made great strides in renewable energy investment and energy efficiency, it has yet to tackle carbon dioxide emissions from fossil fuel use. The job of supplying additional energy to the growing Indian economy is falling on fossil fuels; alternatives are far from keeping pace with the overall demand. In the first decade and a half of this century Indian carbon dioxide emissions have approximately doubled.

This situation highlights a number of issues, with access to energy being a critical driver. Take a country like Botswana, which imports most of its electricity from South Africa — it is also sitting on a coal resource of some 200 billion tonnes,[41] an enormous opportunity that the government is now intent on developing for both domestic use and export. Just keep an eye on the back pages of *The Economist*.

All the above raises one very real question: just how long will it take to change the energy system, and what might that transition look like?

A detailed look[42] at the history of humankind's best efforts for deploying new energy technologies throws up some sobering conclusions. When things have gone well, it has taken around 30 years for a new energy source to move from the first larger-than-laboratory-scale pilot plants, to reach maturity on a material scale. This is defined as delivering 1% of the world's energy supply. Growth follows an exponential curve, running at an average of 26% growth per year for those 30 years — that is, growth has to climb three orders of magnitude in scale during this period.

After that, growth slows down and runs on a more linear course until the energy reaches its ultimate market share in the total energy mix.

This historical evidence is something policymakers need to take seriously before aiming for overly ambitious targets because these global deployment rates are going to be a challenge to beat. Despite the clear observation that solar PV has grown exponentially over the period 2005–2015 and the claims of many that this is evidence of a new pace of transition not seen before, the reality is that solar PV is following a very similar pathway as well. By 2016 this technology was nearing 1% of global primary energy, following its first scaled deployment in the mid-to-late 1980s in California.

Deployment rates based on such findings can be embedded in forward-looking scenarios to help determine how a particular energy source might evolve within the global energy mix. This approach contrasts with what are known as normative scenarios, or ones in which a pathway designed to solve a particular problem is put forward. Many such scenarios exist in climate papers (including those I have helped author for the World Business Council for Sustainable Development).

A normative climate scenario might be one that applies a fixed limit of 1 trillion tonnes for the cumulative release of carbon, and then retrospectively calculates what needs to be done to achieve this, including the required deployment rates for various energy technologies. They are informative, but may not represent what can realistically be delivered. Many apply technically feasible construction rates for deployment of energy technologies and assume huge reductions in energy demand due to efficiency measures, both of which sound feasible, but in practice may not be deliverable. Real world issues such as finance, asset lock-in, depreciation and efficiency rebound,[43] to name but a few, get in the way. New energy technologies may also require new supply chains to be developed or a very significant expansion of an existing supply chain, such as the new demand for certain metals as electric vehicle use grows.

Looking at a modern developed economy today, it is possible to imagine a state of much lower emissions, or even net-zero emissions. The technologies that would enable us to have a zero-emission power sector are readily available and have been for some time — the widespread use of nuclear power in France as early as the 1980s is a good example. Today there is CCS and scalable renewable energy. Vehicle electrification is coming of age, and it is not difficult to see a future where electric vehicles dominate. Heavy transport may use a fuel such as hydrogen, although that being said, John Deere unveiled a fully electric tractor at the Paris International Agribusiness Show in early 2017. Homes can also be electrified, and the service sector/secondary industry economy that drives the developed world today is primarily electricity based.

But the manufacture of goods represents a large part of the global economy, and it is critically important in the early stages of a country's economic development. Turning to a similar but earlier incarnation of Mr Chowdhury in China, a small initial demand led to some 350 million refrigerators being produced and put into domestic use in China between 2004 and 2014, with a further 250 million exported. In 2000 China was producing 13 million refrigerators per annum, but by 2010 this had jumped to 73 million. China is now the world's sixth largest exporter (2014 by value) of refrigerators, although this is just one-sixth of US refrigerator exports.

Development pathways such as those in China and India are placing considerable upward pressure on global emissions, even as regions such as the EU begin to make meaningful reductions. And the level of carbon dioxide in the atmosphere continues to rise.

A SOLAR WORLD

Ending the global use of fossil fuels is widely advocated as a quick solution to meeting the goal of the Paris Agreement. Certainly at the Paris COP and in subsequent forums, renewable energy is touted as the panacea to the problem. More particularly, solar PV is seen by many as the route forward. So could the world be 100% solar, or at least 100% renewables with a largely solar base? The question is often answered in the affirmative by simply demonstrating that so much solar energy falls on the Earth's surface, all energy needs could be met by covering just 500,000 km^2 with solar

PV. This represents an area a bit larger than Thailand, but still only ~0.3% of the total land surface of the planet. Given the space available in deserts in particular and the experience with solar PV in desert regions in places such as California and Nevada, the argument goes that there are no specific hurdles to such an endeavour.

However, there are two significant limitations to solar PV: (1) it is intermittent, and (2) it delivers only electricity, which currently makes up about 20% of final energy use. This is an important point that many energy transition pundits fail to take into account. Final energy is the energy people and businesses use directly to provide goods and services. For example, electricity is used to power lights and devices in our homes, but most of us use gasoline or diesel extracted from crude oil to power our cars. Natural gas is used in many households throughout the world for cooking and to fuel boilers for hot water and heating. Airlines use Jet A1, a kerosene also extracted from crude oil, to power their fleets, and smelters use metallurgical coal to convert an ore to a metal, such as iron in a blast furnace.

In the current energy system, about 80% of final energy is delivered from sources other than electricity. This split is slowly shifting, but it has taken over a century for electricity to reach its current share. The range throughout the world varies from just a few per cent in Gabon to nearly 35% in Sweden and just under 50% in Norway. The global share is currently shifting by 0.25 percentage points per annum in favour of electricity. If that rate doesn't pick up, it will take centuries for electricity to dominate the energy system.

The task confronting any energy transition is therefore twofold: shifting final energy away from point-of-use emitting sources such as oil products and providing sufficient primary energy without emissions to meet all the final energy needs. This helps put the 100% solar energy question into perspective.

The first task in a solar world is of course to generate sufficient electricity, not just in terms of total gigawatt hours, but in gigawatt hours when and where it is needed. As solar is without question intermittent in a given location, this means building a global grid capable of distribution to the extent that any location can be supplied with sufficient electricity from a location that is in daylight at that time. In addition, the same system would likely need access to significant electricity storage, certainly on a scale that far eclipses even the largest pumped water storage currently available. Energy storage technologies such as batteries and molten salt (well suited to concentrated solar thermal) only operate on a very small scale today.

The State Grid Corporation of China has been busy building ultra-high voltage long-distance transmission lines across China. The Chinese have imagined a world linked by a global grid[44] with a significant proportion of electricity needs generated by solar from the equator and wind from the Arctic. But could this idea be expanded to a grid that supplies all the electricity needs of the world? A practical problem here is that for periods of the day at certain times of the year the entire North and South American continents are in complete darkness, which means that the grid connection would have to extend across the Atlantic or Pacific Oceans. While the cost of a solar PV cell may be pennies in this world, the cost of deploying electricity from solar as a global 24/7 energy service could be considerable. The cost of the cells themselves may not even feature.

The idea of solar ambition on such a scale was even discussed by Leonardo DiCaprio in the National Geographic climate change movie *Before the Flood*. The film offers the opportunity to join DiCaprio as he explores the topic of climate change, and discover what might be done today to address it.

DiCaprio tours the world, talking to scientists, business people, Pacific islanders and many others. One riveting conversation is with Elon Musk, founder and CEO of Tesla, recorded as he and DiCaprio tour the Tesla Gigafactory in Nevada. This is a truly gargantuan building, with the largest physical footprint on the planet, designed to build batteries on an immense scale. Musk thinks and talks on an epic scale and in their discussion he tells DiCaprio that just 100 Gigafactories would be sufficient to transition the whole world to sustainable energy. DiCaprio is clearly taken aback by this claim, not with doubt but with the realisation that such a task is actually manageable. I agree with DiCaprio — building 100 Gigafactories seems like an entirely plausible venture. But what would this mean and what impact might it have?

According to Tesla, the Gigafactory will produce 35 GWh of battery capacity per annum. A 2016 model Tesla with a 200–300 mile range has an 85 kWh battery, which means that a Gigafactory can produce about 400,000 such batteries per annum. One hundred Gigafactories would produce at least 40 million such batteries per annum. Given the size of the Gigafactory and allowing for improvements in production efficiency, there is almost certainly upwards room in this estimate. Although Elon Musk made mention of 100 Gigafactories, he doesn't say exactly what capacity per factory he is assuming. But on this basis, and assuming improvements in EV efficiency and battery technology, as well as allowing for smaller city vehicles with smaller batteries, the 100 Gigafactories should be able to supply batteries for all new vehicles in a 100% electric fleet, assuming production of some 80 million new cars per annum.

While this outlook potentially meets all the battery requirements for cars, would that put sufficient storage into the energy system to balance the intermittency requirements for a 100% renewable energy grid, particularly if it is primarily solar? And is this sufficient for all energy, as Elon Musk stated?

Over the course of an energy-system transition to net-zero emissions, the electricity system could potentially grow from some 80–90 EJ in 2016 to 300 EJ,[45] a several fold increase. This results from both a growth in the overall size of the energy system and a shift in electrification from some 20% of final energy to over 50%. Assuming grid connections do not follow the Chinese intercontinental vision, each continental mass would need to manage its own intermittency.

In 2015, North American power generation was in excess of 5000 TWh, but in a highly electrified energy system, demand would likely be significantly higher. Assuming at least 300 million electric cars in North America, up to 20 TWh of storage would be needed — and this level of storage could theoretically power the grid for around a day. Given that over 90% of cars are parked at any particular time, instantaneous vehicle use isn't really a factor; however, the backup requirement would need a significant percentage of parked cars to be physically connected to the grid.

Would this level of battery storage solve the problem of complete solar absence across the Americas for several hours a day in the Northern Hemisphere winter? Assuming some likely support from wind, nuclear and hydroelectricity, together with continental inter-connection, this amount of storage feels at least plausible.

A world of 100 Gigafactories all building batteries for an all-electric fleet and a grid powered by wind and solar, reliably backed up by those same vehicles connected to the grid, is aspirational given where things stand today. It will also require considerably more than just building the Gigafactories. This may become a question about materials and resources, rather than technology and manufacturing capacity.

DEPLOYMENT CONSTRAINTS

In thinking about the energy transition, it is easy to look at the status quo and compare it with a clean energy future of renewables, EVs, bio-polymers and recycling — and then naturally to opt for that future. What is often missing in this thought process is the required change in the global stock of materials to get there.

The early part of 2016 brought great excitement for EV enthusiasts with the announcement of the Tesla Model 3 and the subsequent filling of its order book with over 400,000 vehicles in just a few weeks. With costs coming down and vehicle range improving, there appears to be real consumer interest in EVs, including battery electric, plug-in hybrid and hydrogen fuel cell types.

Over the period 2012−2015 EV growth rates were in the range of 50−100% per annum, but this is quite typical of a new technology with a very small base. As the base increases, year-on-year percentage growth slows sharply, even as absolute production continues to increase. For example, if you have ten cars and buy four more in a given year, your growth in car ownership is 40%. If you then buy five more the following year, your growth in car ownership has fallen to 36%, even though you bought more new cars. Buying six more in the year after that sees car ownership growth fall again to under 32% per annum.

The first major global goal for EV deployment is to reach an installed base of 20 million vehicles by 2020, or about 2% of the global fleet. This is the target set by the Electric Vehicle Initiative of the Clean Energy Ministerial, a global Energy Minister forum to promote policies and share best practices to accelerate the global transition to clean energy. The initiative seeks to facilitate the global deployment of EVs, including plug-in hybrid electric vehicles and fuel cell vehicles.

By the end of 2016 the total global EV stock passed 2 million vehicles,[46] which would give just four years to produce another 18 million cars. That would require year-on-year growth rates of at least 50% per annum into the 2020s in order to produce 1−2 million vehicles per annum and to reach total annual production of 6−7 million vehicles per annum in 2020. Production in 2016 was only about 700,000 vehicles.

If growth at such rates could be achieved and then continue, with additional new production surpassing

4 million per annum throughout the balance of the 2020s and into the 2030s, then by 2035 the global EV stock could be at 500 million vehicles, or nearly a third of the total expected fleet. By this time absolute annual EV growth may be slowing, influenced by an outlook that sees EV production approaching that of global passenger vehicle production. This is assuming that there is no consumer resistance to EVs, even amongst those who love the roar of a finely tuned, high-powered internal combustion engine (ICE).

Should production of EVs completely eclipse that of ICE vehicles, there remains the generational timespan to turn over the entire fleet. The age distribution of vehicles is very broad, so ICE vehicles won't disappear overnight. The average fleet age has also been rising, up from 8.4 to 9.7 years in Europe over the period 2005–2015. There is also a wide distribution; for example, in the Netherlands in 2012, 41% of the passenger vehicle fleet was over 10 years old, but for the same year in Poland, it was 71%.

Putting all the above together in a single very ambitious outlook, ICE vehicles could effectively vanish from the road in about 2060, an endpoint that would be compatible with the goal of the Paris Agreement. This assumes that the global passenger vehicle fleet tops out at around 1.7 billion vehicles in the 2060s, a number which is highly uncertain. For instance, just as EVs are beginning to make progress in the market, autonomous vehicles are possibly offering a completely different model for car ownership, which could see far fewer cars in the global fleet. The prospect of a much smaller market could start to send ripples through the entire investment chain, slowing the uptake of EVs considerably. Equally, if personal motoring progresses rapidly in developing countries, the fleet could be much larger in the second half of the century, which may also argue for an older fleet with ICE vehicles remaining on the road for much longer.

But even in this ambitious scenario, which is predicated on a Paris-compatible outcome, we don't see a decline in ICE vehicles relative to 2015 levels for some time. On the back of overall global fleet growth and existing production, which currently totals over 70 million vehicles per annum, maximum ICE stock isn't reached until well into the 2020s, topping out at about 1.2 billion vehicles vs. 900 million in 2015. ICE numbers return to 2015 levels in the mid-2030s, when the decline really sets in.

Populating the world with 1.7 billion EVs requires building that many batteries and producing the materials that those batteries require. Although we can visualise a world of near 100% recycling of the battery components, the benefits of recycling don't come until much later. Even in a fast-growth scenario the number of batteries being recycled after 15 years use is far lower than the number being produced. For example, in the rapid EV scenario, scrappage of EV vehicles still lags EV production in 2060.

The 2016 Tesla Model S has a nickel-cobalt-aluminium-lithium ion battery. The cathode of the battery consists of about 35 kg nickel and 7 kg cobalt. While there are many different battery chemistry formulations available, each offering different properties in terms of energy density, charging rate, hysteresis, etc., all depend on particular combinations of metals.

In a world in which the above Tesla chemistry dominated, and assuming an eventual global fleet of some

1.7 billion cars (in line with the scenario outlined above), the shift to a 100% EV fleet would require an on-the-road stock build of some 50 million tonnes of nickel and 10 million tonnes of cobalt. This stock is never recovered unless battery chemistry changes or the EV car population falls.

Global production of nickel in 2016 was around 2 million tonnes per annum, with about 3% of that used for batteries and nearly 70% in the stainless-steel industry. Global cobalt production for the same year was around 110,000 tons per annum, with nearly two-thirds of this coming from the Democratic Republic of the Congo. The remainder comes from about ten other countries.

Given these production levels, the required stock build would fully consume 25 years of global nickel production and 91 years of cobalt production. The accelerated EV scenario requires this level of stock build in less than 35 years, which either means a rapid escalation of production in these metals or many different battery chemistry formulations — and probably both. Competition for materials will also come from home batteries, such as the Tesla Powerwall and many other new battery applications, although these may also use alternative formulations.

The initial proposition of the Tesla chemistry dominating isn't simple conjecture either. At an event at COP22 in Marrakech, Professor Jeff Sachs from the Earth Institute at Columbia University made the strong claim that there would be no further production of ICE vehicles after 2030 (a notion that exceeds even the rapid EV scenario given above, where EV and ICE production are about equal in 2030). Scaling up battery production so rapidly would likely depend on an existing chemistry; there simply wouldn't be time to wait for new formulations to be researched, developed and perfected for mass production.

A potential energy transition outcome for transport is one of significantly increased production of certain materials, diversity in battery chemistry, a slower than desired uptake in EVs and perhaps a much smaller EV fleet than anticipated, thanks to autonomous driving and vehicle sharing. Nevertheless, some simple calculations quickly show that the EV transition is likely to be a complex one, with an end result possibly far from the expected.

ENERGY DELIVERY

In relation to the broader energy system transition, although electricity is an important and growing part of the energy mix, it only gets you part of the way there. Different forms of energy will be needed for a variety of processes and services that are unlikely to run on direct or stored electricity, even by the end of this century. For example:

- Shipping currently runs on hydrocarbon fuels, although large military vessels have their own nuclear reactors. Wind could make a return to some extent.

- Aviation requires kerosene, with stored electricity a very unlikely alternative. The fuel-to-weight ratio of electro-chemical (battery) storage, even given advances in battery technology, makes this a distant option. Although a small electric plane for one person for 30 minutes flight has been tested, extending this to an A380 flying for 14 hours would require battery technology that doesn't currently exist. Still, some short-haul commuter aircraft might become electric. Commercial aviation is currently heading towards some 1 billion tonnes of carbon dioxide emitted per annum.

- While electricity may be suitable for many modes of road transport, it may not be as practical for heavy goods transport and large-scale construction equipment. Much will depend on the pace and scope of battery development.

- Heavy industry requires considerable energy input, such as from furnaces powered by coal and natural gas. These reach the very high temperatures necessary for processes such as chemical conversion, making glass, converting limestone to cement and refining ores to metals. Economy of scale is also critical, so delivering very large amounts of energy into a relatively small space is important. In the case of the metallurgical industries, carbon (usually from coal) is also needed as a reducing agent to convert the ore to a refined metal. Electrification will not be an immediately available solution in all cases.

All the above argues for another energy delivery mechanism, potentially helping with (or even solving) the energy storage issue, offering high temperatures for industrial processes and the necessary energy density for transport. The best candidate appears to be hydrogen, which could be made by electrolysis of water in a solar world — although today it is made much more efficiently from natural gas. The resulting carbon dioxide can be geologically stored — an end-to-end process already operating in the Shell Scotford Refinery in Canada. Hydrogen can be transported by pipeline over long distances, stored for a period and combusted directly. Hydrogen could also feature within the domestic utility system, replacing natural gas in pipelines (where suitable) and being used for heating in particular. This may be a more cost-effective route than building sufficient generating capacity and grid reliability to heat homes with electricity on the coldest winter days.

Hydrogen could play a major role in the industrial space, both as a fuel for processes such as cement manufacture and even as the reducing agent in metallurgical processes. The latter has been researched and patents awarded for a single-step process for the preparation of high purity iron by using hydrogen plasma in a suitable smelting reactor furnace. But the process has not been commercialised or even demonstrated at scale.

The scale of a global hydrogen industry to support a solar world would far exceed the global Liquefied Natural Gas (LNG) industry that exists in 2017. That sector includes around 300 million tonnes per annum of liquefaction capacity and some 400 LNG tankers. That amounts to about 15 EJ of final energy compared to the current global primary energy demand of 500 EJ. In a 1000 EJ world that could exist in 2100, a role for hydrogen as an energy carrier that reached 100 EJ would require an industry that was seven times the size of the current LNG system. But hydrogen has 2–3 times the energy content of natural gas, and liquid hydrogen is one-sixth the density of LNG (important for ships), so a very different looking industry would emerge. Nevertheless, the scale would be substantial.

Finally, but importantly, there are the things that we use, from plastic water bottles to the Tesla Model S. Everything has carbon somewhere in the supply chain or in the product itself. There is simply no escaping this. The source of carbon in plastics, in the components in a Tesla and in the carbon fibre panels of a Boeing 787 is crude oil (and sometimes natural gas). So our solar world needs a source of carbon and on a very large scale. This could still come from crude oil, but if one objective of the solar world were to contain that genie, then an alternative would be required. Biomass is one and a bioplastics industry already exists. In 2015 it was $1-2$ million tonnes per annum, compared to ~350 million tonnes for the traditional petroleum-based plastics industry.

Another source of carbon could be from carbon dioxide removed directly from the atmosphere or sourced from industries such as cement manufacture. With sufficient energy input this can be converted to synthesis gas (carbon monoxide and hydrogen) or combined with hydrogen to make methane, both of which can be precursors for the chemical industry or an ongoing liquid fuels industry for sectors such as aviation and shipping. This route forward also returns us to the hydrogen story.

SECTOR CHALLENGES

I first stepped onto a Boeing 747 in 1971 and in 1979 I flew to London for the first time on a Qantas 747. In early 2015 I flew to San Francisco on a Virgin 747, albeit a slightly longer, more sophisticated, efficient and larger capacity one than the 1980 model — but still a 747 burning many tons of jet fuel. In 1990, the US Presidential Air Force One 707 was replaced by a 747, which still flies today, but which the USAF announced in 2015 could be replaced with the 747−8 in the early 2020s. Those planes are likely to fly for some 30 years which means that one version or another of the 747 will have been visible in the skies for a period of more than 80 years. This is also true for other planes being built today, with many just entering the beginning of their production runs (787, A350, A380), rather than heading towards the end, as might be seen with the 747 series.

Although planes will certainly evolve over the course of the century, the rate of change is likely to be slow, particularly if a step change in technology is needed. In 100 years of civil aviation there have been two such step changes: the first commercial flights in the 1910s and the shift of the jet engine from the military to the commercial world with the development of the Comet and Boeing 707. The 787 Dreamliner is in many respects a world away from the 707, but in terms of the fuel used, it is the same plane. That's 60 years — and there is little sign of the next change.

The challenge for aviation is to find a fuel or energy carrier of sufficiently high energy density that it is capable of flying a modern jet aeroplane. Direct use of hydrogen is a possibility, but the resulting change in the fuel-to-volume ratio could mean a radical redesign of the whole shape of the plane as well as possibly requiring the redesign of infrastructure such as airport terminals, air bridges and so on. Even the development and first deployment of the double decker A380, something of a step change in terms of shape and size, has taken 20 years and cost Airbus and airports many billions.

Therefore, the simplest approach will probably be the development of a process to produce a lookalike hydrocarbon fuel. The most practical way to approach this problem is via an advanced biofuel route, and a few processes are already available, although scale-up of these technologies has yet to take place. But what if the biofuel route also proves problematic — say, for reasons related to land use change or perhaps public acceptance in a future period of rising food prices? A synthetic fuel route is entirely possible from a chemistry perspective, but it requires energy — at least as much energy as the finished fuel gives when it is used and its molecules are returned to water and carbon dioxide.

Audi, amongst others, has been working on such a project. In 2015 they announced the production of the first fuel from their pilot plant (160 litres per day). According to their media release the plant requires carbon dioxide, water and electricity as raw materials. The carbon dioxide is extracted from the ambient air using direct air capture. In a separate process, an electrolysis unit splits water into hydrogen and oxygen. The hydrogen is then reacted with the carbon dioxide in two chemical processes conducted at 220°C and a pressure of 25 bar to produce a hydrocarbon liquid, which is called Blue Crude. This conversion process is up to 70% efficient and runs on solar power.

Apart from the front end of the facility with its air capture and electrolysis, the rest of the plant should be very similar to the full-scale gas-to-liquids (GTL) facilities that Shell and SASOL each operate in Qatar. Here natural gas is converted to synthesis gas, which is, in turn, converted to a mix of longer-chain hydrocarbons, including jet fuel (contained within the Audi Blue Crude). The Shell GTL facility is capable of producing 140,000 barrels per day of products, so perhaps 100 such facilities would be required to produce enough jet fuel for the world (this would depend on the yield of suitable jet fuel from the process, which produces a range of hydrocarbon products that can be put to many uses). Today there are just a handful of GTL plants in operation across the entire oil and gas industry — two in Qatar, one in Malaysia and two in South Africa. The final conversion uses the Fischer-Tropsch process, originally developed about a century ago. This process coverts the synthesis gas into hydrocarbon liquids.

Each of these future Blue Crude facilities would also need a formidable solar array to power it. The calorific content of the finished fuels is about 45 TJ/kt — the minimum amount of energy required for the conversion facility. However, accounting for the efficiency of the process and adding in the energy required for air extraction of carbon dioxide and all the other energy needs of a modern industrial facility, a future process might need up to 100 TJ/kt of energy input. The Shell GTL facility produces 19 kt of product per day, so the energy demand to make this from water and carbon dioxide would be 1900 TJ per day, or 700,000 TJ per annum. As such, this requires a nameplate capacity for a solar PV farm of about 60 GW — roughly equal to a fifth of the entire installed global solar PV generating capacity in 2017. A Middle East location such as Qatar receives about 2,200 kWh/m^2 per annum, or 0.00792 TJ/m^2. Assuming a future solar PV facility that might operate at 35% efficiency (considerably better than commercial facilities today), the solar PV alone would occupy an area of some 250 km^2 — which is 500 km^2 or more in total plot area for the facility.

While such a large-sized facility is not inconceivable, it is far larger than any solar PV facility in operation today. For example, in 2015, the largest solar plant in the world, the Topaz solar array in California, occupied a site 25 square kms in size with a nameplate capacity of 550 MW. It produced about 1.1 million MWh per annum (4000 TJ), but with an efficiency (23%) far lower than the 35% efficiency of the future assumption mentioned earlier. At this production rate, 175 Topaz farms would be required to power a refinery with the hydrocarbon output of the Shell GTL facility. My assumptions represent a packing density of solar PV that is four times more efficient than Topaz (i.e. 100 MW/km^2 vs. 22 MW/km.^{2})

To summarise, this means that our (non-biofuels) solar world might need to see the construction of at least 100 large-scale hydrocarbon synthesis plants, together with air extraction facilities, energy storage for night time operation of the reactors and huge solar arrays. This could meet all future aviation needs and be capable of producing lighter and heavier hydrocarbons for various other applications where electricity is not an easy option (e.g. chemical feedstock and heavy marine fuels). The investment would certainly run into trillions of dollars and take decades to implement.

The aviation industry has recognised these challenges and at least for the coming decade or more has settled on a more commercially viable approach. In the midst of the global rush to see the Paris Agreement enter into force, the International Civil Aviation Organisation (ICAO) settled on a global Market Based Measure (MBM) system to begin to manage the emissions of carbon dioxide from international aviation. The IEA 2014 assessment for the international aviation sector (excluding domestic aviation which is covered under NDCs) puts emissions at 504 million tonnes of carbon dioxide or 1.5% of global combustion emissions from energy use. This represents a rise of 95% over 1990 levels versus a 58% rise in carbon dioxide emissions more broadly for the same period. Nevertheless, this rise indicates a considerable efficiency improvement for aviation, as Revenue Passenger Kilometres (RPK) had nearly tripled in the same period. A September 2016 market update from Boeing noted that China alone could order some 7,000 planes to a value of $1 trillion through to the mid-2030s, which implies continued growth in aviation emissions at a level well in excess of energy system emissions.

As was the case with the Kyoto Protocol, international aviation emissions are not covered by the Paris Agreement. Rather, the aviation industry has spent the years between Copenhagen and Paris negotiating an industry agreement to manage emissions. While there is an important emphasis on efficiency improvements and the longer-term application of synthetic fuels, primarily from biomass, the centrepiece of the approach is an offset system utilising the global carbon markets. Collectively, the approach will be known as the Carbon Offsetting and Reduction Scheme for International Aviation (CORSIA). This is similar to an approach that the IMO were considering for the marine sector in the run-up to COP15 in Copenhagen, but didn't implement.

The CORSIA calls for international aviation to address and offset its emissions through the reduction of emissions elsewhere (outside of the international aviation sector), involving the use of 'emissions units'. Two main types of emissions units exist: 'offset credits' from crediting mechanisms and 'allowances' from emissions-trading schemes. Offsetting could be through the acquisition and surrender of emissions units, arising from different sources of emission reductions achieved through mechanisms (e.g. UNFCCC's Clean Development Mechanism), programmes (e.g. REDD + — reducing emissions from deforestation and forest degradation in developing countries) or projects (e.g. substituting coal-fired stoves with solar cookers). The CORSIA will be implemented in a phased approach throughout the 2020s.

While the offset approach envisaged for the 2020s will use the full range of current emissions-trading instruments, as emissions fall globally, aviation will need to look to a sequestration only strategy — i.e. 'sinks', to use the terminology of the Paris Agreement. ICAO has hinted at this with the inclusion of forestry in the portfolio of measures, but a long-term reliance on nature-based offsets may not be sufficient to meet their needs. They will also need to consider early investment in nascent offset technologies such as BECCS and DACCS. While industrial CCS has emerged as a scalable technology, offshoots such as DACCS operate at pilot scale. Given sufficient attention from ICAO, these could be at a demonstration level in the 2020s, with the goal of scalability by 2030.

The offset approach adopted by the aviation industry fits with the direction of the Paris Agreement in that it seeks to balance out the emission sources rather than find zero-emission alternatives to fuel the sector. This approach is likely to be replicated in other sectors where the engineering challenge and cost of alternative energy sources make the use of emission sinks such as direct CCS, BECCS and even DACCS a more attractive proposition. The use of CCS on a very large scale therefore becomes a major component in the delivery of net-zero emissions in the second half of the century. The technology is already proving very effective today, despite the small scale on which it operates.

RETURNING TO CCS

One of the many report launches at COP22 was the 2016 Global Status of CCS, released by GCCSI. The report identified 15 large-scale CCS projects in operation around the world, with a carbon dioxide capture capacity of close to 30 million tonnes per annum (Mtpa). With additional projects coming on line in Australia and North America in particular, the number of large-scale operational CCS projects is expected to increase to 21 by the end of 2017, with a capture capacity of approximately 40 Mtpa.

Encouraging as this is, it is not nearly enough. The report notes that over the decade 2005–2015 CCS investment has totalled $20 billion, compared with $2.5 trillion for clean energy investment. But in many instances, the justification for clean energy investment is claimed to be for low emissions and consequent climate benefits, rather than simply for the energy generated. If that is the case, how do these two approaches compare on a pure climate basis?

CCS is a technology that is entered into almost entirely for climate reasons. While some carbon dioxide is used for Enhanced Oil Recovery (EOR) and various niche applications, for the most part, this is a technology designed to prevent carbon dioxide from entering the atmosphere when fossil fuels are used, instead returning it to the geosphere. By contrast, a technology such as solar PV is designed to produce electricity, which may then displace a certain fossil fuel usage that might have been used to generate the same amount of electricity. However, its effectiveness for climate mitigation purposes depends on the nature of the displacement — the displaced fuel could be used elsewhere or consumed later, which could negate some or all of the benefit claimed.

Also by 2017 there will be some 300 GW of solar PV in the world. The latter has had at least an order of magnitude more fiscal support than that offered to CCS, and according to GCCSI, clean energy in total has had over two orders of magnitude more investment than CCS. Let's assume that a third of this is solar, so $800 billion.

To assess the climate benefit of solar PV, let us use a solar company figure; the First Solar project, Desert Sunlight Solar Farm, is 550 MW and according to the project website displaces some 300,000 tonnes of carbon dioxide annually. On that basis 300 GW of solar is displacing 163 Mtpa of carbon dioxide, compared with the 40 Mtpa that CCS is achieving.

So for ~$30 billion ($20 billion in the period 2006–2015 and perhaps ~$10 billion prior to 2006), CCS is achieving a quarter the climate benefit as ~$800 billion of solar, on an investment basis. Of course, this is not the complete story as the CCS incurs an annual operating cost and the solar PV provides electricity, but there is a factor of nearly seven here in terms of carbon dioxide benefit in favour of the CCS.

The GCCSI report highlights the effectiveness of CCS when faced with the issue of emissions mitigation, which is what the climate issue is really all about. Simply building out renewable energy capacity with the hope of permanently displacing fossil fuel use may not be a cost-effective approach when climate is the single objective. Of course, we don't live in a world of single objectives, but nevertheless the calculation above demonstrates that CCS needs to be given a much greater opportunity to flourish and do the job for which it is ideally positioned — permanently removing carbon dioxide.

REAL WORLD SCALING

Can global energy sources be reorganised in just a few decades? There is no doubt that to solve the climate issue quickly and decisively and have a chance of keeping the surface temperature rise well below 2°C, this needs to happen — but that doesn't mean it can. There has to be tremendous political will to do so, but political will isn't enough to overturn the existing industrial capacity that society relies on today, let alone replace it with a set of technologies that in some instances don't yet exist. The development and deployment of radical new technologies can take decades.

The stretch goal within the Paris Agreement is to limit warming of the climate system to just 1.5°C. At the time of COP21, Dr. Joeri Rogelj was one of the few researchers who had studied 1.5°C pathways. His analysis showed that such a pathway required net-zero carbon dioxide emissions by around 2050 and net-negative emissions, in other words a drawdown of carbon dioxide in the atmosphere, from 2050 to 2100. This raises the clear question of whether it is possible for net-zero emissions to be achieved within 35 years after the Paris Agreement. While 35 years may sound like enough time, to give you some perspective, 35 years is the average span of a person's career. In fact, it represents my career in the oil and gas industry from when I began work with Shell in Australia in 1980 to the time of the Paris Agreement, where you will find me at Shell in London. In terms of some technologies, a lot has changed. In 1980 there were no personal computers in Geelong Refinery; today it probably couldn't run without them, although the distillers, crackers and oil movement facilities have hardly changed and in some instances are precisely the same pieces of equipment that were running in 1980.

An energy transition spanning a single career is one that would have to grapple not just with the birth of a range of new industries to deal with the emissions in a number of key sectors of the economy, but also with the construction of sufficient non-emitting primary energy capacity.

Scaling up is a formidable challenge, but it is something that industry does very well — given sufficient impetus. One of the best examples of an industrial scale push was the production of Liberty ships in the United States during the Second World War. These were cargo ships built for the dangerous trans-Atlantic run when U-Boats were an ever-present hazard. During the period 1941–1945, 18 American shipyards built 2,751

vessels — easily the largest number of ships ever produced to a single design. The ships were constructed
in assembly-line style, using sections that were then
welded together. This was a new technique requiring
new skills. Early on the ships took the best part of a
year to build, but the average length of construction
time eventually dropped to 42 days (with a much
heralded record of 4 days and 15½ hours for the
Robert E. Peary).

In 1943, three new Liberty ships were being completed
every day. Production of Liberty ships saw step gains in
productivity and capacity as new shipyards opened. In
1941 production was at about 100−150 ships per annum
(although real production didn't start until the second
part of the year), 300−400 per annum in 1942 and
1100 per annum in 1943 and beyond (until production
pretty much ended in 1945). In less than three years
Liberty ship capacity in the United States increased by a
factor of nearly 10.

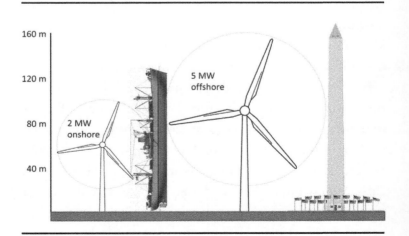

A modern-day clean-energy comparison for the United States might be the production of wind turbines. By 2015 US-installed wind capacity was approaching 70 GW, with growth averaging about 8 GW per annum over the period 2010–2015, about 50% above the rate seen in 2005–2010. Manufacturing capacity in the United States has grown rapidly due to foreign and domestic investment. Nacelles, towers, blades and other components are not made by the same manufacturers or even necessarily in the United States, but on balance, US manufacturing capacity has increased several fold over the 21st century. Not, however, at the rate of Liberty ships.

Given current ambitions for clean energy deployment and the potential for 200 GW of installed wind capacity by the early 2020s, further opportunities exist. A step change on the scale of Liberty ship capacity could see the United States not only meeting its own demand for turbines, but also becoming an exporter of equipment. If a Liberty ship could be equated to, say, 10 MW of installed wind capacity (two large turbines over 100 metres high vs. 4,000 tons of steel in a 135-metre-long Liberty ship), then maximum wartime capacity of Liberty ships might approximate to manufacturing 2500 + 'Liberty Turbines' per annum, enough to reach the installed capacity objective.

There is a parallel story today with the current global scale-up of LNG capacity. This was a technology that first appeared in the 1960s and saw a scale-up over the 1970s and 1980s to 60 million tonnes per annum globally. As energy demand soared in the 1990s and 2000s, LNG production quickly rose again to around 300 million tonnes per annum in 2016 and could reach 500 million tonnes per annum by 2030.

CCS is also a critical technology that needs to see global scale-up if society is serious about achieving net-zero emissions in the latter part of the century. There are many parallels between LNG production and CCS that may offer some insight into the potential for CCS. Both require drilling, site preparation, pipelines, gas-processing facilities, compression and gas transport, although LNG also includes a major cryogenic[47] step, which is not necessary for CCS.

LNG production and CCS are both gas-processing technologies; the comparison between them needs to be in terms of volume, not tonnes. Carbon dioxide has a higher molecular weight than natural gas (methane), so the processing of a million tonnes of natural gas is the same as nearly 3 million tonnes of carbon dioxide. As such, the production scale up to 500 million tonnes of LNG by 2030 could be equated to nearly 1.5 billion tonnes of carbon dioxide per annum in CCS terms, which is a number that starts to be significant in terms of real mitigation. The actual scale-up from 2015 to 2030 is projected to be 200–250 million tonnes of LNG, which in CCS terms is about 700 million tonnes per annum of carbon dioxide.

The scale-up of LNG shows that industrial expansion of a complex process involving multiple disciplines from across the oil and gas industry is entirely possible. LNG took two to three decades to reach 100 million tonnes, but less than 10 years to repeat this level of growth. In the following 10 years (2010–2020) production should nearly double again with an additional 200 million tonnes of capacity added. These latter rates of scale up are what is needed now for technologies such as CCS, but the world is clearly languishing in the early stages of deployment, with just a few million tonnes of production (if that) being added each year.

What is missing in support of CCS is the strong commercial impetus that LNG has seen over the last 15 years as global energy demand has shot up. With most, if not all, of the technologies needed for CCS widely available in the oil and gas industry, it may be possible to shorten the initial early deployment stage that can last 20 years (as it did for LNG). If this shortening could be achieved, CCS deployment at rates of a billion tonnes per decade, for starters, may be possible. This is the minimum scale needed for mitigation that will make a tangible difference to the task ahead.

The Second World War provided the powerful policy driver for Liberty ship development and deployment. Today, by comparison, energy policy is fragmented and uncertain. The full potential of the global economy to deliver change has not been realised. Some people have argued that society will need to go on to a 'war footing' for there to be real progress.

8

REACHING AN OUTCOME

An uninformed belief is gaining ground and leading to the false conclusion that a very rapid energy transition is underway that will solve the emissions issue. This belief is that renewable energy is becoming so cost-competitive that emissions will fall rapidly and decisively without real financial outlay, other than the natural cost of energy system replacement.

As already shown, this scenario is unlikely. While there is no doubt that an energy transition is in the making, the situation later in this century may not be one of net-zero emissions if such wishful thinking about renewable energy prevails. Rather, an energy mix could evolve that is very different from the current one, but that still gives rise to a level of emissions that is unacceptably high, continuing to drive further warming of the climate system.

The IEA shows 2014 final energy demand at around 360 EJ (with an additional 40 EJ for non-energy use, e.g. petrochemicals), with electricity at 73 EJ, or about 20%. Biofuels and waste contribute around 50 EJ. If final energy demand approached 800 EJ later this century, with 400 EJ supplied as electricity from renewables and biofuels doubling to 100 EJ, the remaining demand would be 300 EJ, similar to current levels. Global carbon dioxide emissions would certainly be lower as the power generation system is renewable-energy based, but they could still be as high as 20 Gt, compared with 32 Gt in 2014.

Given a 2014 population of around 7.2 billion, final energy use per capita was around 55 GJ per annum, although OECD energy use was 120 GJ per capita. A late century population of 10 billion consuming energy at two-thirds of current OECD levels gives an 800 EJ final energy demand. This is why efficiency is so important over the long term: 10 billion consuming at current OECD levels would mean a tripling of the global energy system.

Despite an ongoing energy transition, emissions management needs to become a critical activity throughout this century. That means a particular focus on carbon pricing policies and recognition that CCS will be an essential technology for decades to come, also but one that needs much greater impetus as a result of Paris. The role that CCS needs to play continues to be undervalued and even disregarded.

Within the energy transition, the global community hopes to find an alternative energy pathway forward for the 3 billion people on the planet who currently lack or have little access to energy, while seeking to decarbonise the emerging and developed countries that have already built or are well advanced in building their energy systems. This is the thinking behind the Paris Agreement. With this in mind, the NDCs put forward by the less developed economies are particularly important as they potentially hold the key to avoiding a 3+°C world in 2100. But these submissions all have a critical dependency, the availability of money to ply an alternate pathway, which doesn't include large-scale use of indigenous resources such as coal.

FINDING THE MONEY

One such pre-Paris NDC offers the opportunity for a closer look at this issue.

The Kenya NDC proposes a 30% reduction in national greenhouse-gas emissions from a business-as-usual trajectory. The plan notes that Kenya is aiming to be a newly industrialised middle-income country by 2030. Current emissions are very low, with the majority coming from land-use change. In 2010 emissions were 73 Mt in carbon dioxide terms, with the IEA reporting carbon dioxide emission from energy use at 11.4 Mt for that year. Given the population of 41 million in 2010, that gives energy-linked carbon dioxide emissions per capita of 0.28 tonnes, amongst the lowest in the world. Kenya has projected that its greenhouse gas emissions on a business-as-usual trajectory will reach the equivalent of 143 million tonnes of carbon dioxide equivalent by 2030, so that gives them a goal of 100 million tonnes for that year on the basis of their intended contribution.

Kenya has also made it clear that their NDC is subject to international support in the form of finance, investment, technology development and transfer, and capacity building. With some of this support coming from domestic sources, Kenyan officials estimate the total cost of mitigation and adaptation actions across sectors at US$40 billion, through to 2030. These numbers are all open to challenge, but they help frame the issue and allow some assessment to establish a ballpark estimate of value for money and the implications flowing from that.

- Let's assume that the $40 billion is split between mitigation and adaptation, but with emphasis on mitigation. That allows ~$10 + billion for major public works and capacity-building programmes focussed on areas such as water and agriculture and $20–$30 billion in the energy system.

- I will assume that energy system growth and adaptation funding allow for a plateau and then gradual decline in land-use change emissions, such that by 2050 these are below 10 MT per annum.

- A continuation of business-as-usual for energy emissions only would see Kenya rising to nearly 2 tonnes per capita by 2030 (current Asia, excluding China) and 6 tonnes per capita by 2050 (approaching current Europe). This would mean extensive use of fossil fuels, but supplemented by geothermal and hydroelectric resources in particular. This is the pathway that they might be on in the absence of this NDC.

- Kenya's population rises in line with the UN mid-level scenario, i.e. to 66 million by 2030 and 97 million by 2050.

Based on the above, energy emissions could rise to some 120 Mt p.a. by 2030 and 600 Mt p.a. by 2050 under this business-as-usual scenario. But in the NDC scenario, this could be curtailed to 70 Mt p.a. in 2030 and perhaps as low as 130 Mt p.a. in 2050, or 70–80% below business-as-usual. The 2030 number is the critical one for this calculation as this is what the $20–$30 billion delivers, although the benefits of the investment stretch beyond 2030. However, further additional investment would be required to keep emissions at such a low level through to 2050 as energy demand grows.

The deviation from our business-as-usual scenario is nearly 50 Mt p.a. by 2030, with that deviation starting in the early 2020s. If the gains are held through to 2050, then the cumulative emission reduction over the period is around 1 billion tonnes. On a simple 20-year project life with no discounting, that equates to around $25 per tonne of carbon dioxide against the $20–$30 billion investment in the 2020s. On that basis, as a mixture of expanded renewable energy deployment, natural gas instead of coal and possibly some biofuel development for transport, this is well within the bounds of plausibility. Within a broad distribution of mitigation activities where some are well below $25 per tonne of carbon dioxide, it might even see some industrial CCS deployment at the upper end of the range.

What is perhaps more interesting is how this scales up across Africa and other parts of the world where energy access is currently limited. If 3 billion people globally need support for similar energy infrastructure, that implies a financial requirement of about US$2 trillion over the period 2020–2030 just for mitigation — that is, the global call is at least 60 times that of Kenya with its population of 50 million, so 60 multiplied by $US30 billion. This financial requirement is the equivalent of $200 billion per annum over the 10-year period.

Despite 2009's Copenhagen agreement around the need for wealthier nations to channel climate finance and direct support to the developing economies over the 10-year period from 2010 to 2020 (by which point it should reach $100 billion per annum), it was never clear how this financial flow would be measured. Nevertheless, the $100 billion has stood as a benchmark number for support although its form has been contested ever since. Some have argued that it represents the amount that should be deposited annually in the Green Climate Fund. The more widely held view is that it represents a score-card financial assessment of the collective mechanisms and approaches for funding and financing low-carbon energy systems and adaptation programmes in developing countries. In 2015 an OECD report established that the level of support being provided from developed country governments, development banks such as the World Bank and private sector institutions had reached some $62 billion per annum. This was before the Green Climate Fund had even pledged funding to its first project.

Returning to the NDC calculation above, the starting point is the new reference to $100 billion per annum, now appearing in the decision text supporting the Paris Agreement. Rather than being a goal to strive for, it represents a minimum or floor amount from which to build. The parties are instructed to agree the new amount by 2025, which could then be the $200 billion calculated above.

It also implies that if the world does build on the US$100 billion per annum floor, then most of this will be for mitigation in the least developed economies as they build their 21st century energy systems. The flip side of this funding being directed to the least developed economies is that the emerging economies will have to self-fund, which argues for the implementation of a carbon price on a far wider basis than is currently envisaged. China is leading the way here, but so, too, are countries like Mexico and Chile.

The Kenya NDC offers some interesting insight into climate politics in the years to come. The successful implementation of the Paris Agreement depends, at least in part, on the issue of climate finance.

However, the possibility remains that even if the net-zero emissions goal is achieved, cumulative emissions will still build to an unacceptably high level. That possibility should lead to a very different debate, one that is unwelcome by many but that is, nevertheless, the subject of ongoing, albeit low-key, research.

GEOENGINEERING

In the days before computer models, climate lobbyists, climate sceptics, so-called warmists and catastrophists, there was an early and thoughtful introduction to and analysis of the issue of climate change by the science community, published within a longer report on the environment by the US government and issued under the signature of President Lyndon B. Johnson. It makes for fascinating reading.

The 1965 paper looked at the atmospheric build-up of carbon dioxide, the potential for further build-up by 2000 as fossil fuels continued to be consumed, expected temperature rises and the possible impact this temperature rise would have on global sea levels as ice caps melted. It concluded that:

> .... *the climatic changes that may be produced by the increased carbon dioxide content could be deleterious from the point of view of human beings.*

Perhaps the most surprising aspect of the 1965 paper is the reference to a geoengineering solution, the deliberate large-scale intervention in the Earth's natural systems to counteract climate change. The conclusion to the section on carbon dioxide emissions goes on to say;

> *The possibilities of deliberately bringing about countervailing climatic changes therefore need to be thoroughly explored. A change in the radiation balance in the opposite direction to that which might result from the increase of atmospheric carbon dioxide could be produced by raising the albedo, or reflectivity, of the earth.......*

Nearly 50 years later, not a great deal has been done in response to this, although climate science has certainly advanced. Rather, the focus has been entirely on mitigation of emissions, with an increasing parallel focus on adaptation. The case for taking more drastic action through geoengineering has waned, even as success in actually reducing emissions has been hard to come by.

There are good reasons for this. The climate system is chaotic, at least in the short term. Climate science is still struggling to fully understand and forecast shorter decadal trends, such as the apparent pause in surface temperature change in the first decade of this century. Although the physics of a geoengineering solution may be understood in terms of the radiative balance, the shorter-term impact could be unpredictable and therefore potentially dangerous.

Nevertheless, some scientific effort continues against the backdrop of political steps such as the Paris Agreement. The simplest of the geoengineering solutions is already in existence, albeit not as an intentional means of managing the global temperature. Sulphur is being artificially pumped into the troposphere through the world-wide use of High Sulphur Fuel Oil (HSFO) in ships (and of course from other sources such as coal-fired power stations not fitted with scrubbers). The combustion of this fuel powers much of the world's oceangoing fleet, with the sulphur being emitted through the ship's funnel. HSFO contains some 3.5% sulphur, so a modern container ship travelling from Shanghai to Southampton via the Suez Canal will eject about 30 tonnes of sulphur into the atmosphere, along with some 3,000 tonnes of carbon dioxide. The carbon dioxide adds to the growing accumulation of this gas in the atmosphere, but the sulphur remains in the atmosphere for just a few weeks in aerosol form before dropping out. Nevertheless, as a result of all the marine activity and other sources of sulphur, there is a net suspension of sulphur in the atmosphere above us. The result of this is that it cools the atmosphere by scattering incoming radiation, offsetting some of the warming impact of carbon dioxide and other greenhouse gases.

However, sulphur also has a negative effect in terms of local and regional air quality so the International Maritime Organization (IMO) has moved to limit sulphur in marine fuel. One analysis[48] discusses the climate impact of the marine fuel sulphur specification being reduced globally to

0.5%. Whereas the global annual average cooling effect of shipping is currently some -0.6 W/m^2 (compared to the current additional radiative forcing from post-industrial carbon dioxide now around 2 W/m^2), this is shown to reduce to -0.3 W/m^2 in the case of a global 0.5% sulphur specification — in other words, another 0.3 W/m^2 of warming.

But the real issue is the potential role of sulphur in deliberately managing the global temperature as a possible geoengineering solution. Trying to do this at sea level and injecting sulphur into the troposphere has far less impact than doing the same in the stratosphere. For the same amount of surface cooling, approximately one-twentieth the amount of sulphur is required at 25,000 metres because the half-life of the aerosol suspension is some 18 months at that height, rather than a few weeks seen in the low atmosphere.

An indicative calculation has shown that a fleet of 150 aircraft injecting sulphur into the stratosphere on a continuous basis could potentially offset the warming associated with a doubling of carbon dioxide in the atmosphere. The cost of this is estimated to be no more than $10 billion per annum and perhaps quite a bit less. But what of the implications of being able to manage atmospheric warming for an amount so small that even some individuals could undertake the experiment, or perhaps a group such as the small island states in defence of their territory? For major emitters this would be a paltry sum, far less than some of the direct mitigation options. But if such a practice were undertaken, what then for the global endeavours to reduce emissions? Would we all give up trying? And while some amount of cooling might be achieved, phenomena such as ocean acidification would continue. Who should decide on such weighty issues and what if one nation or group of nations decided to conduct the practice unilaterally?

There are other geoengineering options, but perhaps none as simple as the sulphur solution. At the other end of the spectrum are proposals to reflect sunlight before it reaches the Earth. Suggested shade designs include a single-piece shade and a shade made by a great number of small objects. Most such proposals contemplate a blocking element at the Sun-Earth L1 Lagrange point.

A more predictable geoengineering solution might be one that reverses the cause of our warming climate and begins to remove carbon dioxide from the atmosphere. As previously discussed, the application of CCS and other NETs will be critical.

IN CONCLUSION

The Paris Agreement has sent a signal around the world: climate change is a serious issue that governments across the globe are determined to address. But the world is rapidly developing, with energy demand continuing to rise. Several billion people still aspire to a first refrigerator, as did Mr Chowdhury in India. Others are looking towards a first car or perhaps a long-haul flight to visit other parts of the world. These activities are very likely to put upward pressure on carbon dioxide emissions, even as renewable energy is deployed rapidly throughout the power generation sector and as electric mobility becomes more commonplace. Society will respond to these new demands and seek to meet them, as has been the case for the entire Industrial Revolution.

The climate deal that has been carefully and tirelessly negotiated over many years is not a solution in itself — rather, it is a roadmap to help all of us find a way forward. The effort that was put into the Paris Agreement now has to be taken up by every cabinet office, every parliament and every national institution across the world if the Agreement is to cascade effectively throughout the global economy. The success of the Paris Agreement will also require extraordinary transparency, governance and institutional capacity. Businesses must respond as well, but their behaviour is primarily driven by the market — hence the need for governments to introduce carbon-pricing measures.

By the turn of this century, but ideally before, there is the potential for a very different energy system to emerge. It can be a system that brings modern energy services to all in the world, without delivering a climate legacy that society cannot readily adapt to. That is the essence of the Paris Agreement.[49]

NOTES

[1] Many geologists have noted that humankind is having such an impact on the planet that it is ushering in a new climatic era, taking over from the Holocene.

[2] Emissions from human activities such as the combustion of fossil fuels, rather than natural emissions as part of the carbon cycle.

[3] Group Climate Change Adviser, located in the Health, Safety and Environment team in the Royal Dutch/Shell Group Corporate Centre as it existed in 2001.

[4] The first global deal between nations to try to limit the emissions of greenhouse gases such as carbon dioxide to the atmosphere.

[5] Chlorofluorocarbons, used as refrigerants for most of the second half of the 20th century.

[6] Natural gas found in remote locations is compressed and cooled to a liquid for shipment to major energy use hubs such as Japan, China and Europe.

[7] These are polyethylene (PE), polypropylene (PP), polyvinyl chloride (PVC), polystyrene solid (PS), polyethylene terephthalate (PET) and polyurethane (PUR).

[8] A big coal-fired power station will be at least 1 GW (gigawatt).

[9] Daniel J. Jacob (1999), *Introduction to atmospheric chemistry*, Princeton University Press.

[10] When a tonne of carbon is combusted it will form 3.667 tonnes of carbon dioxide, the additional weight coming from the oxygen atoms that are now combined with the carbon.

[11] Allen et al. (2009, 30 April). Warming caused by cumulative carbon emissions towards the trillionth tonne. *Nature, 458.*

[12] The paper expressed the problem in tonnes of carbon as this is more easily equated to the fossil resource itself (e.g. as coal is largely carbon), but it needs to be multiplied by 3.667 to give tonnes of carbon dioxide as actually emitted to the atmosphere when the fossil fuel is used.

[13] For example, the 2013 paper from Scripps Institution of Oceanography (The role of HFCs in mitigating 21st century climate change, Y. Xu1, D. Zaelke, G. J. M. Velders, and V. Ramanathan).

[14] A measure of dispersion in a frequency distribution.

[15] Probabilistic assessment of sea level during the last interglacial stage. Kopp, R. E., Simons, F. J., Mitrovica, J. X., Maloof, A. C., & Oppenheimer, M. (2009). Probabilistic assessment of sea level during the last interglacial stage. *Nature, 462,* 863–867.

[16] Pfeffer, W., Harper, J. T., & O'Neel, S. (2008). Kinematic constraints on glacier contributions to 21st-century sea-level rise. *Science, 321*(5894), 1340.

[17] http://trillionthtonne.org/

[18] Meinshausen, & Allen et al. (2009, 30 April). Greenhouse-gas emission targets for limiting global warming to 2°C. *Nature, 458,* 1158–1162.

[19] Information reference document. Prepared and adopted by EU Climate Change Expert Group 'EG Science'. Final Version, Version 9.1, 9 July 2008, 16:15.

[20] James Hansen is Adjunct Professor in the Department of Earth and Environmental Sciences at Columbia University. Hansen is best known for his testimony on climate change to congressional committees in 1988. From 1981 to 2013, he was Head of the NASA Goddard Institute for Space Studies.

[21] Turn Down the Heat: Why a 4°C World Must Be Avoided.

[22] Analysis of climate policy targets under uncertainty. Webster, M., Sokolov, A.P., Reilly, J.M. et al. *Climatic Change*, (2012) *112*, 569. doi:10.1007/s10584-011-0260-0

[23] Climate expert says China is a decade ahead of schedule on reducing CO_2 emissions. *Forbes*, July 2016.

[24] 2016 BP Statistical Review of World Energy.

[25] Carbon dioxide equivalent, i.e. including other greenhouse gases such as methane.

[26] Recovery of additional oil from a reserve beyond that originally anticipated when the extraction started. A typical reserve may only yield half the oil that is actually there, but this can be increased with enhanced recovery methods applied later in the life of the field.

[27] GCCSI database, July 2014.

[28] One tonne of bituminous coal contains about 0.7 tonnes carbon (2.6 tonnes CO_2), but one tonne of lignite may hold only 0.4 tonnes carbon (1.5 tonnes CO_2). Using 0.5 (1.8 tonnes CO_2), the total cost of the coal becomes $50 + 1.8 \times 75 = \$187$ per tonne.

[29] Direct action carbon reduction policy running out of steam (2016). *The Guardian,* November 24.

[30] The Hartwell Paper, A new direction for climate policy after the crash of 2009. Prins et al. Institute for Science, Innovation and Society, University of Oxford. MacKinder Programme, London School of Economics.

[31] Biophysical and economic limits to negative carbon dioxide emissions (2016). *Nature Climate, 6.*

[32] "The New Lens Scenarios" and "A Better Life with a Healthy Planet" are part of an ongoing process — scenario-building — used in Shell for more than 40 years to challenge executives' perspectives on the future business environment. Shell bases them on plausible assumptions and quantification, and they are designed to stretch management thinking and even to consider events that may only be remotely possible. Scenarios, therefore, are not intended to be predictions of likely future events or outcomes, and investors should not rely on them when making an investment decision with regard to Royal Dutch Shell plc securities.

It is important to note that Shell's existing portfolio has been decades in development. While Shell believes the portfolio is resilient under a wide range of outlooks, including the IEA's 450 scenario, it includes assets across a spectrum of energy intensities including some with above-average intensity. While Shell seeks to enhance its operations' average energy intensity through both the development of new projects and divestments, Shell has no immediate plans to move to a net-zero emissions portfolio over their investment horizon of 10–20 years.

[33] Paltsev, S., Sokolov, A., Chen, H., Gao, X., Schlosser, A., Monier, E., … Haigh, M. (2016). Scenarios

of global change: Integrated assessment of climate impacts. *Joint Program Report Series*, February, 34 p.

[34] Energy [r]evolution — A sustainable world energy outlook 2015.

[35] Paltsev, S., Reilly, J., & Sokolov, A. (2013). What GHG concentration targets are reachable in this century? *Joint Program Report Series*, p. 9.

[36] Down to around 100 metres.

[37] 'Climate Related Death of Coral Alarms Scientists' (2016). *New York Times*, April 10.

[38] When a company, organisation or individual uses their resources to obtain an economic gain from others without reciprocating a reasonable benefit back to society through wealth creation.

[39] Reducing emissions from deforestation and forest degradation (REDD) is a mechanism that has been under negotiation by the UNFCCC, with the objective of mitigating climate change through reducing net emissions of greenhouse gases through enhanced forest management in developing countries.

[40] Institute for Energy Research, India Opening Coal Mines; Will Surpass U.S. in Coal Production, October 12, 2015.

[41] Department of Energy, Government of Botswana.

[42] G.J. Kramer and M. Haigh (2009, 3 December). No quick switch to low carbon energy. *Nature, 462*

[43] The increased use of an energy service due to its lower cost, e.g. air conditioning, or the use of the same energy for other services due to its release from the initial use it was put to as a result of improved efficiency.

[44] *Wall Street Journal*, 30 March 2016 and *Bloomberg*, 3 April 2016.

[45] A Better Life with a Healthy Planet: Pathways to Net-Zero Emissions. Royal Dutch Shell, 2016.

[46] Electric cars set to pass 2m landmark globally by end of 2016, *The Guardian*, October 2016.

[47] Cooling the gas to a very low temperature, $-160°C$ in the case of LNG.

[48] James, J. W., et al. (2009). Assessment of near-future policy instruments for oceangoing shipping: Impact on atmospheric aerosol burdens and the earth's radiation budget, *Environmental Science & Technology*, 43(15).

[49] Cautionary Note: The companies in which Royal Dutch Shell plc directly and indirectly owns investments are separate legal entities. In this book 'Shell', 'Shell group' and 'Royal Dutch Shell' are sometimes used for convenience where references are made to Royal Dutch Shell plc and its subsidiaries in general. Likewise, the words 'we', 'us' and 'our' are also used to refer to subsidiaries in general or to those who work for them. These expressions are also used where no useful purpose is served by identifying the particular company or companies. "Subsidiaries", 'Shell subsidiaries' and 'Shell companies' as used in this book refer to companies over which Royal Dutch Shell plc either directly or indirectly has control. Entities and unincorporated arrangements over which Shell has joint control are generally referred to as 'joint ventures' and 'joint operations' respectively. Entities over which Shell has significant influence but neither control nor jointly control are referred to as 'associates'. The term 'Shell interest' is used for convenience to indicate the direct and/or indirect ownership interest held

by Shell in a venture, partnership or company, after exclusion of all third-party interest.

This book contains forward-looking statements concerning the financial condition, results of operations and businesses of Royal Dutch Shell. All statements other than statements of historical fact are, or may be deemed to be, forward-looking statements. Forward-looking statements are statements of future expectations that are based on management's current expectations and assumptions and involve known and unknown risks and uncertainties that could cause actual results, performance or events to differ materially from those expressed or implied in these statements. Forward-looking statements include, amongst other things, statements concerning the potential exposure of Royal Dutch Shell to market risks and statements expressing management's expectations, beliefs, estimates, forecasts, projections and assumptions. These forward-looking statements are identified by their use of terms and phrases such as 'anticipate', 'believe', 'could', 'estimate', 'expect', 'goals', 'intend', 'may', 'objectives', 'outlook', 'plan', 'probably', 'project', 'risks', "schedule", 'seek', 'should', 'target', 'will' and similar terms and phrases. There are a number of factors that could affect the future operations of Royal Dutch Shell and could cause those results to differ materially from those expressed in the forward-looking statements included in this book, including (without limitation): (a) price fluctuations in crude oil and natural gas; (b) changes in demand for Shell's products; (c) currency fluctuations; (d) drilling and production results; (e) reserves estimates; (f) loss of market share and industry competition; (g) environmental and physical risks; (h) risks associated with the identification of suitable potential acquisition properties and targets, and successful negotiation and completion of such transactions; (i) the risk of doing business in developing countries and countries subject to international sanctions; (j) legislative, fiscal and regulatory developments including regulatory measures addressing climate change; (k) economic and financial market conditions in various countries and regions; (l) political risks, including the risks of expropriation and renegotiation of the terms of contracts with governmental entities, delays or advancements in the approval of projects and delays in the reimbursement for shared costs; and (m) changes in trading conditions. There can be no assurance that future dividend payments will match or exceed previous dividend payments. All forward-looking statements contained in this book are expressly qualified in their entirety by the cautionary statements contained or referred to in this end note. Readers should not place undue reliance on forward-looking statements. Additional risk factors that may affect future results are contained in Royal Dutch Shell's 20-F for the year ended 31 December 2016 (available at www.shell.com/investor and www.sec.gov). These risk factors also expressly qualify all forward-looking statements contained in this book and should be considered by the reader. Each forward-looking statement speaks only as of the date of publication of this book. Neither Royal Dutch Shell plc nor any of its subsidiaries undertake any obligation to publicly update or revise any forward-looking statement as a result of new information, future events or other information. In light of these risks, results could differ materially from those stated, implied or inferred from the forward-looking statements contained in this book.

The author may have used certain terms, such as 'resources', in this book that US Securities and Exchange Commission (SEC) strictly prohibits Shell from including in its filings with the SEC. US investors are urged to consider closely the disclosure in the Shell Form 20-F, File No 1-32575, available on the SEC website www.sec.gov

INDEX